T0325095

A Guide to the Collision Avoidance Rules

A Guide to the Collision Avoidance Rules
International Regulations for Preventing Collisions at Sea

Seventh edition

Incorporating the 1981, 1987, 1989, 1993, 2001 and 2007

Amendments

A. N. Cockcroft and J. N. F. Lameijer

AMSTERDAM • BOSTON • HEIDELBERG • LONDON • NEW YORK
OXFORD • PARIS • SAN DIEGO • SAN FRANCISCO • SINGAPORE
SYDNEY • TOKYO
Butterworth-Heinemann is an imprint of Elsevier

Butterworth-Heinemann is an imprint of Elsevier
The Boulevard, Langford Lane, Kidlington, Oxford, OX5 1GB, UK
225 Wyman Street, Waltham, MA 02451, USA

First published by Stanford Maritime Ltd 1965
Second edition 1976
Reprinted 1978
Third edition 1982
Reprinted 1984, 1985, 1987
Fourth edition 1990
Reprinted 1990, 1991
Revised and reprinted 1993
Fifth edition 1996
Reprinted 1997, 1998, 1999 (twice), 2000, 2001
Sixth edition 2004
Seventh edition 2012

British Library Cataloguing in Publication Data
A catalogue record for this book is available from the British Library

Library of Congress Cataloguing in Publication Data
A catalog record for this book is available from the Library of Congress

ISBN 978-0-08-097170-4

For information on all Butterworth-Heinemann publications
visit our website at www.elsevierdirect.com

Printed and bound in Great Britain by MPG Books Ltd, Bodmin, Cornwall

The authors and publishers, whilst exercising the greatest care in compiling this book, do
not hold themselves responsible for the consequences arising from any inaccuracies therein.

Contents

Preface

When the Regulations for Preventing Collisions at Sea were revised at an International Conference held in 1960 the changes made were generally of minor character. Soon after the amended Regulations came into force in 1965 it became apparent that a more thorough revision was necessary to take account of such developments as the widespread acceptance and use of radar, the introduction of traffic separation and the increase in size and speed of many ships. At an International Conference held in October 1972 substantial changes were made and a new format was adopted. The 1972 Regulations came into force in 1977.

This book contains the complete text of the 1972 Regulations together with an explanation of the changes which have been made and background information about the intentions of the International Conference. A number of coloured illustrations have been included to show the new arrangements of lights and signals and there are also several diagrams to explain certain aspects of the Steering and Sailing Rules.

Although major changes have been made to the Rules much of the original wording is still incorporated. Some of the phrases have been considered many times in the Courts and extracts from Court judgments which remain relevant in the context of the 1972 Regulations have been included in this book to show how the Rules have been interpreted.

In recent years much information has become available on the manœuvring characteristics of ships. The stopping distances and turning circles of ships of various types and sizes are shown in a number of diagrams at the end of the book. The effects of interaction are also described and illustrated. Recommendations on manœuvres to avoid collision are included together with a manœuvring diagram.

A sound knowledge of the Regulations for Preventing Collisions at Sea has always been considered to be essential for navigating officers. Candidates for examinations have sometimes been encouraged to commit the Rules to memory but this does not necessarily result in a clear understanding of the contents. The purpose of this book is to promote a better understanding of the Rules by discussing the implications of the various phrases and giving Court interpretations.

The authors are grateful for assistance received from a number of people in the preparation of this book, and particularly to Captain A. C. Manson, who was chairman of the IMCO Working Group, for commenting on the text.

Preface to the Seventh Edition

The seventh edition incorporates amendments to Annex IV of the Collision Regulations adopted by the 25th Assembly of the International Maritime Organization in 2007 and Unified Interpretations of Annex I, adopted by the Maritime Safety Committee of IMO in 2008.

Some new collision cases have been added and some have been deleted as considered no longer relevant to present day circumstances. Some of the early extracts from court judgments have been retained as they still appear to be applicable.

Previous editions of this book have included comparisons with former rules and explanation of changes made by the 1960 Conference. These have been deleted as they have become less relevant.

The opportunity has been taken to review, and make some changes to, the comments on the Rules.

Collisions and the Courts

Reporting a collision

If a United Kingdom ship becomes involved in a collision a report must be made by the master and sent to the Chief Inspector of Marine Accidents or made directly to a Marine Office of the Department of Transport or to HM Coastguard. When a ship is lost such a report must be made by the owner, master or senior surviving officer.

Preservation of evidence

The owner and master must, so far as is possible, ensure that all charts, log books and other records and documents which might reasonably be considered pertinent to a collision be kept and no alteration made to entries therein. Any equipment which might be considered pertinent to a collision must so far as is practicable be left undisturbed. The above requirements apply until notification is received that no investigation is to take place or until the inspector carrying out the investigation no longer requires such evidence.

Investigation of a collision

If a collision occurs within UK territorial waters or if a UK ship is involved in a collision elsewhere or possibly in certain other circumstances the Marine Accident Investigation Branch (MAIB) may hold an investigation. The extent of the investigation depends upon the circumstances. In some minor cases the Chief Inspector may decide that the initial report provides enough information and that no further action is needed. For more serious accidents there is likely to be an investigation by an Inspector, which in a major case may take the form of an Inspector's Inquiry. If an Inspector's Inquiry is ordered public notice is given and written representations are invited.

The purpose of an investigation is to determine the circumstances and causes of the collision with the aim of improving the safety of life at sea and the avoidance of future accidents. The purpose is not to apportion liability or blame. Following an investigation the Chief Inspector submits a report to the Secretary of State. The draft is sent to any parties who are criticized and any representations they make are considered before the report is finalized. The report may be published and the Secretary of State must order publication if the report relates to a serious casualty to a UK ship or if it appears that to do so will improve safety of life at sea and help to prevent accidents in the future.

Formal investigation

The Secretary of State for Transport may decide that a formal investigation should be held into the circumstances and causes of a collision, conducted by a wreck commissioner assisted by one or more assessors. The wreck commissioner is a person of wide experience in maritime law; the assessors in collision cases are usually certificated masters with at least two years' experience in command and a wide knowledge of modern aids to navigation. The formal investigation will normally be held in public.

The main purpose of a formal investigation is to determine the cause of the collision in the interests of safety of life at sea but a charge may be made against individuals if this may help to bring about the avoidance of future casualties. All parties to the investigation may be represented by counsel. The Attorney General and other parties may produce witnesses who may be examined, cross-examined and recalled if necessary. After the examination of witnesses all parties may address the wreck commissioner upon the evidence.

The wreck commissioner has the power to cancel or suspend the certificates of ships' officers, and may in addition, order the parties concerned to contribute to the costs of the investigation.

'Reports of Courts', concerning formal investigations are published by Her Majesty's Stationery Office and are put on sale to the public.

Admiralty Courts

After a collision has occurred an action may be brought to recover damages. An action may be brought in the country of cither the plaintiffs or the defendants, or in any other country where the law permits such actions to be brought. The Courts in the United Kingdom and the United States will allow an action to be brought if the defendant vessel is in a port of their country at the time.

In the United Kingdom an action for damages after a collision would be held in a Court of Admiralty jurisdiction, usually the Admiralty Court in London. The proceedings there would be held before a High Court Judge who is usually assisted by two of the Elder Brethren of Trinity House acting as nautical assessors to give advice on matters of seamanship.

Appeals from the Admiralty Courts are heard by the Court of Appeal. Three Lords Justices are usually present and they may be assisted by two nautical assessors. No witnesses are called.

If leave is granted a further appeal may be made to the House of Lords. Such an appeal would be heard before five Law Lords, usually assisted by two assessors.

History of the Collision Regulations

For several hundred years there have been rules in existence for the purpose of pre-venting collisions at sea, but there were no rules of statutory force until the last century. In 1840 the London Trinity House drew up a set of regulations which were enacted in Parliament in 1846. One of these required a steam vessel passing another vessel in a narrow channel to leave the other on her own port hand. The other regulation relating to steam ships required steam vessels on different courses, crossing so as to involve risk of collision, to alter course to starboard so as to pass on the port side of each other. There were also regulations for vessels under sail including a rule, established in the eighteenth century, requiring a sailing vessel on the port tack to give way to a sailing vessel on the starboard tack.

The two Trinity House rules for steam vessels were combined into a single rule and included in the Steam Navigation Act of 1846. Admiralty regulations con-cerning lights were included in this statute two years later. Steam ships were required to carry green and red sidelights as well as a white masthead light. In 1858 coloured sidelights were prescribed for sailing vessels and fog signals were required to be given, by steam vessels on the whistle and by sailing vessels on the fog horn or bell.

A completely new set of rules drawn up by the British Board of Trade, in consultation with the French Government, came into operation in 1863. By the end of 1864 these regulations, known as Articles, had been adopted by over thirty maritime countries including the United States and Germany.

Several important regulations which are still in force were introduced at that time. When steam vessels were crossing so as to involve risk of collision the vessel with the other on her own starboard side was required to keep out of the way. Steam vessels meeting end-on or nearly end-on were required to alter course to starboard. Every vessel overtaking any other had to keep out of the way of the vessel being overtaken. Where by any of the rules one vessel was to keep out of the way the other was required to keep her course.

In 1867 Mr Thomas Gray, Assistant Secretary to the Marine Department of the Board of Trade, wrote a pamphlet on the Rule of the Road at Sea which included a number of verses as aids to memory. The verses became popular and were translated into other languages. They are still quoted in some text books.

Some changes to the 1863 Rules were brought into force in 1880, including a new rule permitting whistle signals to be given to indicate action taken by steam ships to avoid collision. In 1884 a new set of regulations came into force but these did not differ substantially from the previous ones. An article specifying signals to be used by vessels in distress was added bringing the total number of articles to 27.

The first International Maritime Conference to consider regulations for preventing collision at sea was held in Washington in 1889. It was convened on the initiation of the Government of the United States of America. Among the new provisions agreed at the Conference were requirements that a stand-on vessel should keep her speed as well as her course, that a giving-way vessel should avoid crossing ahead of the other vessel, and that steamships should be permitted to carry a second white masthead light.

The regulations agreed at the Washington Conference were brought into force by several countries, including Britain and the United States, in 1897. At a further Maritime Conference held in Brussels in 1910 international agreement was reached on a set of regulations which differed in only minor respects from those drafted at the Washington Conference. The 1910 Regulations remained in force until 1954.

In 1929 an International Conference on Safety of Life at Sea proposed some minor changes to the Rules but these were never ratified. However, a recommendation that helm and steering orders should be given in the direct sense, so that 'right rudder' or 'starboard' meant 'put the vessel's rudder to starboard', was accepted and came into force in 1933. The situation with respect to helm orders had previously been confused due to the difference between the movement of the wheel and tiller.

The Regulations were revised at an International Conference on Safety of Life at Sea in 1948. No drastic changes were made. The second masthead light was made compulsory for power-driven vessels of 150 feet or upwards in length, a fixed stern light was made compulsory for almost all vessels under way, and the wake-up signal of at least five short and rapid blasts was introduced as an optional signal for use by a stand-on vessel. The revised Rules came into force in 1954.

Relatively few vessels were fitted with radar in 1948 so no changes were made to take account of this equipment. However, the Conference did add a recommendation that possession of a radio navigational aid in no way relieves a master of a ship from his obligations under the International Regulations and under Rules 15 and 16 (applying to vessels in restricted visibility) in particular.

With the considerable increase in the number of ships fitted with radar during the following years, coupled with a series of collisions involving such vessels, it became apparent that further revision of the Rules was necessary. An International Conference on Safety of Life at Sea was convened in London in 1960 by the Inter-Governmental Maritime Consultative Organization (IMCO), which is now the International Maritime Organization (IMO).

At the 1960 Conference it was agreed that a new paragraph should be added to the Rules governing the conduct of vessels in restricted visibility to permit early and substantial action to be taken to avoid a close quarters situation with a vessel detected forward of the beam. Recommendations concerning the use of radar were made in an Annex to the Rules. The changes were not confined to the Rules relating to restricted visibility but most of the other amendments were relatively minor in character. These Rules came into force in 1965.

In September 1960 the British Institute of Navigation set up a working group to consider the organisation of traffic in the Dover Strait. The French and German

Institutes of Navigation agreed to co-operate in the following year and a separation scheme was devised. A new working group with representatives from additional countries was formed in 1964 to consider routeing schemes for other areas. The proposals were accepted by IMCO and recommended for use by mariners in 1967.

An International Conference was convened in London in 1972 by IMCO to consider the revision of the Regulations. The Conference agreed to change the format so that the Rules governing conduct preceded the Rules concerning lights, shapes and sound signals. Technical details relating to lights, shapes and sound signals were transferred to Annexes. The Stand-on Rule was amended to permit action to be taken at an earlier stage and more emphasis was placed on starboard helm action in both clear and restricted visibility. New Rules were introduced to deal specifically with look-out requirements, safe speed, risk of collision and traffic separation schemes.

International Regulations for Preventing Collisions at Sea, 1972 (with comments)

Part A – General

RULE 1

Application

(a) These Rules shall apply to all vessels upon the high seas and in all waters connected therewith navigable by seagoing vessels.

(b) Nothing in these Rules shall interfere with the operation of special rules made by an appropriate authority for roadsteads, harbours, rivers, lakes or inland waterways connected with the high seas and navigable by seagoing vessels. Such special rules shall conform as closely as possible to these Rules.

(c) Nothing in these Rules shall interfere with the operation of any special rules made by the Government of any State with respect to additional station or signal lights, shapes or whistle signals for ships of war and vessels proceeding under convoy, or with respect to additional station or signal lights, or shapes for fishing vessels engaged in fishing as a fleet. These additional station or signal lights, shapes or whistle signals shall, so far as possible, be such that they cannot be mistaken for any light, shape or signal authorized elsewhere under these Rules.

(d) Traffic separation schemes may be adopted by the Organization for the purpose of these Rules.

(e) Whenever the Government concerned shall have determined that a vessel of special construction or purpose cannot comply fully with the provisions of any of these Rules with respect to the number, position, range or arc of visibility of lights or shapes, as well as to the disposition and characteristics of sound-signalling appliances, such vessel shall comply with such other provisions in regard to the number, position, range or arc of visibility of lights or shapes, as well as to the disposition and characteristics of sound-signalling appliances, as her Government shall have determined to be the closest possible compliance with these Rules in respect of that vessel.

COMMENT:

(a) Seaplanes, non displacement craft and WIG craft are included in the definition of a vessel given in Rule 3(a) and should therefore be considered as power-driven vessels for the purpose of these Rules, except as specifically provided for in Rules 18(e), 18(f) and 31.

(b) Roadsteads are included among the areas for which an appropriate authority may lay down special rules. A roadstead is an open anchorage, generally protected by shoals, which offers less protection than a harbour. This means that coastal states or local authorities may make special rules for areas which lie outside the usual

A Guide to the Collision Avoidance Rules. DOI: 10.1016/B978-0-08-097170-4.00001-5

limits of inland waters. Mariners should be aware that special rules may be applicable in such areas and should consult the Sailing Directions and other publications for details.

Special rules concerning lights, sound signals and other aspects of collision avoidance are in force in many ports, rivers, canals and inland waters throughout the world.

The final sentence of paragraph (b) is included to stress the need for conformity. It is hoped that there will be no proliferation of special rules and that the authorities will make every effort to eliminate any important differences with the 1972 Regulations which would be likely to confuse mariners.

(c) This Rule refers to additional lights and signals for fishing vessels, warships, etc., authorised by the Government of any State and does not apply to the signals for vessels fishing in close proximity listed in Annex II which have received international agreement. Paragraph (c) was amended in 1981 to include shape(s).

(d) This section of the Rule gives IMO (the Organization) the authority to adopt traffic separation schemes to which the provisions of Rule 10 of the 1972 Regulations will apply. A traffic separation scheme is defined by IMO as follows: 'A routeing measure aimed at the separation of opposing streams of traffic by appropriate means and by the establishment of traffic lanes.' A traffic separation scheme is a routeing measure, the particulars of which are laid down in the IMO publication 'Ships' Routeing'. This publication gives details of all traffic schemes adopted by the Organization. Amendments are issued to enable the publication to be kept up to date and information about new schemes, or amendments to existing schemes, are also promulgated through Notices to Mariners. It is important to keep nautical charts and publications updated with respect to any changes concerning traffic separation schemes.

(e) This paragraph originally had particular application to naval vessels but was also applied to other vessels of special construction and purpose, when full compliance with the provisions of the Rules for lights, shapes or sound-signalling appliances could not be achieved without interfering with the special function of the vessel.

When in 1986 the data expired for exemptions relating to provisions mentioned in paragraphs (d)(ii), (e), (f) and (g) of Rule 38, a number of governments were of the opinion that repositioning or refitting of lights and/or sound signal appliances as a consequence of the above mentioned requirements would be too onerous for ships flying their flags and not really necessary for the improvement of safety at sea. After due consideration of this problem in IMO it was decided to make the application of Rule 1(e) more general by deleting the phrase – 'without interfering with the special function of the vessel' – in the text of this Rule.

Warships' lights

The special arrangements of lights on some British warships are described in *The Mariner's Handbook*. Aircraft carriers have their masthead lights placed off the centre line with reduced horizontal separation. Their sidelights may be on either side of the hull or on either side of the island structure. Many warships of over 50 metres in length cannot be fitted with a second masthead light.

Submarines usually have two masthead lights but the forward white light may be lower than the sidelights. Some submarines are fitted with an amber flashing light 2 metres above the after masthead light for use as an aid to identification in narrow waters and areas of dense traffic. A similar light is used by hovercraft (see page 109).

RULE 2

Responsibility

(a) **Nothing in these Rules shall exonerate any vessel, or the owner, master or crew thereof, from the consequences of any neglect to comply with these Rules or of the neglect of any precaution which may be required by the ordinary practice of seamen, or by the special circumstances of the case.**

(b) **In construing and complying with these Rules due regard shall be had to all dangers of navigation and collision and to any special circumstances, including the limitations of the vessels involved, which may make a departure from these Rules necessary to avoid immediate danger.**

COMMENT:

(a) Precautions required by good seamanship or special circumstances

Some examples of precautions which may be required by the ordinary practice of seamen, or by special circumstances, are the following:

1. A vessel under way would be expected to keep clear of a vessel at anchor as a matter of seamanship. But a vessel under way and stopped must not rely on other vessels keeping out of her way, unless she is not under command and is displaying the appropriate signals; she must comply with the Rules.

2. When a vessel anchors she must do so without endangering other vessels which may be navigating close by. She must not anchor too close to other anchored vessels. Sufficient cable must be put out according to circumstances and a second anchor should be used if necessary.

3. In dense fog a vessel without operational radar may not be justified in being under way at all but should anchor if it is safe and practicable for her to do so.

4. When two vessels are approaching one another at a difficult bend in a tidal river it has been held to be the duty of the one having the tide against her to wait until the other has passed (see page 67–68).

5. The effects of shallow water must be taken into account. A vessel moving at fairly high speed through the water produces pressure fields which become much greater when the flow of water around the ship is restricted. There is a reduction of pressure beneath the ship which causes bodily sinkage so that the vessel is said to 'squat' in the water. In addition to an increase in mean draught there will usually be a change of trim, by the bow or stern according to the circumstances. When the depth of water is less than about one and a half times the draught this effect is much more pronounced. If there is shallow water on only one side the pressure fields may cause the ship to sheer away from the bank which could bring danger of collision if another vessel is passing close by. Interaction between ships due to the pressure fields will also be greater in shallow water and the steering qualities are likely to be affected (see pages 71–73).

(b) The term 'vessels' includes non-displacement craft, WIG craft and seaplanes by the definition of Rule 3.

Dangers of navigation and collision

A departure from the Rules may be required due to dangers of navigation or to dangers of collision. For instance, a power-driven vessel meeting another power-driven vessel end on may be unable to alter her course to starboard, as directed by Rule 14, owing to the presence of shallow water close by to starboard or to the fact that a third vessel is overtaking her on her starboard side.

Special circumstances and immediate danger

This Rule does not give any vessel the right to take action contrary to the Regulations whenever it is considered to be advantageous to do so. A departure is only permitted when there are special circumstances and there is immediate danger. Both conditions must apply. The departure must be of such a nature as to avoid the danger which threatens.

Squadrons or convoys

The Mariners' Handbook draws the attention of mariners to the dangers which may be caused by single vessels attempting to pass ahead of, or through a squadron of warships or merchant vessels in convoy. Single vessels are advised to take early measures to keep out of the way, and the vessels in the squadron or convoy are warned to keep a careful watch and be ready to take such action as will best aid to avert collision.

Mariners are expected to take into account the cautions and recommendations given in Notices to Mariners and other official publications but if a vessel in a formation or convoy is approached by a single vessel so as to involve risk of collision the Steering and Sailing Rules must be complied with.

Action taken in accordance with the advice to avoid a squadron or convoy on the port bow would not be a departure from the Rules if executed at long range before risk of collision begins to apply (see pages 26–28).

Duty to depart if necessary

If a departure from the Rules is necessary to avoid immediate danger a vessel would not only be justified in departing from them but may be expected to do so.

Tasmania-City of Corinth

It is provided by Rule ... that, in obeying and construing the Rules, due regard shall be had to any special circumstances which may render a departure from them necessary in order to avoid immediate danger. As soon then as it was, or ought, to a master of reasonable skill and prudence, to have been obvious that to keep his course would involve immediate danger, it was no longer the duty of the master of the Tasmania *to adhere to the ... Rule. He was not only justified in departing from it, but bound to do so, and to exercise his best judgement to avoid the danger which threatened. (Lord Herschell, 1890)*

RULE 3

General Definitions

For the purpose of these Rules, except where the context otherwise requires:

(a) The word 'vessel' includes every description of water craft, including non-displacement craft, WIG craft and seaplanes, used or capable of being used as a means of transportation on water.

(b) The term 'power-driven vessel' means any vessel propelled by machinery.

(c) The term 'sailing vessel' means any vessel under sail provided that propelling machinery, if fitted, is not being used.

(d) The term 'vessel engaged in fishing' means any vessel fishing with nets, lines, trawls or other fishing apparatus which restrict manoeuvrability, but does not include a vessel fishing with trolling lines or other fishing apparatus which do not restrict manoeuvrability.

(e) The word 'seaplane' includes any aircraft designed to manoeuvre on the water.

(f) The term 'vessel not under command' means a vessel which through some exceptional circumstance is unable to manoeuvre as required by these Rules and is therefore unable to keep out of the way of another vessel.

(g) The term 'vessel restricted in her ability to manoeuvre' means a vessel which from the nature of her work is restricted in her ability to manoeuvre as required by these Rules and is therefore unable to keep out of the way of another vessel.

   The term 'vessel restricted in her ability to manoeuvre' shall include but not be limited to:

(i) a vessel engaged in laying, servicing or picking up a navigation mark, submarine cable or pipeline;

(ii) a vessel engaged in dredging, surveying or underwater operations;

(iii) a vessel engaged in replenishment or transferring persons, provisions or cargo while underway;

(iv) a vessel engaged in the launching or recovery of aircraft;

(v) a vessel engaged in mine clearance operations;

(vi) a vessel engaged in a towing operation such as severely restricts the towing vessel and her tow in their ability to deviate from their course.

(h) The term 'vessel constrained by her draught' means a power-driven vessel which because of her draught in relation to the available depth and width of navigable water is severely restricted in her ability to deviate from the course she is following.

(i) The word 'underway' means that a vessel is not at anchor, or made fast to the shore, or aground.

(j) The words 'length' and 'breadth' of a vessel mean her length overall and greatest breadth.

(k) Vessels shall be deemed to be in sight of one another only when one can be observed visually from the other.

(l) The term 'restricted visibility' means any condition in which visibility is restricted by fog, mist, falling snow, heavy rainstorms, sandstorms or any other similar causes.

(m) The term 'wing-in-ground (WIG) craft' means a multimodal craft which, in its main operational mode, flies in close proximity to the surface by utilizing surface-effect action.

COMMENT:

The definitions given in Rule 3 are those which have general applications throughout the Rules. Definitions concerning lights and whistle signals are given in sections C and D (Rules 21 and 32).

(a) and (b) Non-displacement craft, WIG craft and seaplanes are to be considered as power-driven vessels by the definitions of Rules 3(a) and 3(b); specific responsibilities of WIG craft and seaplanes are set out in Rules 18(e) and 18(f). Rule 31 relates to the lights and shapes to be exhibited by seaplanes and WIG craft.

(c) It will be apparent from the context of the Rules that a vessel propelled by machinery which is 'not under command' or 'restricted in her ability to manœuvre' or 'engaged in fishing' is not always to be regarded as a 'power-driven vessel'. For instance a power-driven vessel engaged in trawling must not show the two masthead lights specified by Rule 23, nor give the sound signals prescribed in Rule 35(a) and (b), and is not required to keep out of the way of a sailing vessel or a 'power-driven vessel' on her starboard side which is crossing so as to involve risk of collision. However, all power-driven vessels, including hampered vessels, would probably be expected to have their engines ready for immediate manœuvre in restricted visibility (Rule 19(b)) and to alter course to starboard when meeting a power-driven vessel of the same category end-on (Rule 14).

(d) The phrase 'which restrict manœuvrability' in the definition of 'vessel engaged in fishing' makes it clear vessels fishing with a few short lines or other small gear, not appreciably affecting their ability to manœuvre, are not entitled to the degree of privilege allocated to vessels engaged in fishing by Rule 18, and must not show the lights and shapes prescribed in Rule 26.

(e) Non-displacement craft are not to be considered as seaplanes for the purpose of Rule 18(e).

(f) For a vessel to be considered 'not under command' there must be some exceptional circumstances, such as breakdown of steering gear, or loss of propulsion power, resulting in a restriction of her ability to manœuvre as required by the Rules of Part B of the Steering and Sailing Rules. Vessels engaged in difficult towing operations and other vessels which from the nature of their work are unable to manœuvre as required by the Rules are now to be considered as 'vessels restricted in their ability to manœuvre'. Such vessels are given the same degree of privilege as vessels not under command but they show different lights and shapes.

It could be argued that adverse weather conditions are not really exceptional and that a vessel would not, therefore, be justified in showing not under command signals when unable to manœuvre in rough seas.

At the 1972 Conference it was considered that adverse weather conditions seriously affecting a vessel's ability to manœuvre would be exceptional circumstances. However, the fact that a vessel's ability to manœuvre is affected by weather conditions does not necessarily mean that she is not under command. The conditions must be so exceptional, with respect to the particular vessel, as to render her unable to keep out of the way of another vessel by alteration of course and/or speed in order to justify the showing of not under command signals.

In addition to vessels which have had a breakdown of engines or steering gear, or which have lost a propeller or rudder, examples of vessels which are likely to be accepted as being not under command under the 1972 Rules are: a vessel with her anchor down but not holding, a vessel riding to anchor chains with anchors unshackled, and a sailing vessel becalmed.

In the case of *Glamorgan–P. Caland,* 1893, it was held in the House of Lords that the *P. Caland* was not justified in exhibiting the not under command lights. The speed of the *P. Caland* had been reduced from 11 knots to about 4 to 5 knots by an accident to the machinery. The *Glamorgan,* seeing the red lights but not the side lights of the *P. Caland,* steamed towards her to offer assistance and collided with her. Lord Herschel, the Lord Chancellor, said:

Under these circumstances I cannot hold that, owing to the disablement of the machinery, the risk of its ceasing to work was so imminent that the vessel can be said not to have been under command within the meaning of the Rule.

Mendip Range–Drake

If a vessel is in such a condition owing to an accident that she can only get out of the way of another after great and unusual delay, I think she must be considered as 'not under command' for the purpose of Article … She is not able to behave as those on board other vessels meeting her would reasonably expect. (Viscount Finlay, 1921)

In 1969 a collision occurred in the Dover Strait during bad weather conditions between the *Ziemia* and the *Djerada.* The *Djerada,* proceeding at 6 ½ knots, was exhibiting not under command lights. Mr Justice Brandon said:

There is no doubt that the Djerada *had been in some difficulties because of the heavy weather but it seems to me to be difficult to say that she was even partly*

disabled. She had the full use of her engines and steering ... I think that the Djerada
was well able to keep out of the way of the other ship without great or unusual delay
and that she had no business to advertise herself as unable to do so.

(g) A definition to cover vessels engaged in operations which restrict manœuvrability
is necessary as such vessels form one of the categories referred to in Rule 18 which
specifies responsibility for keeping out of the way.

The lead-in sentence to sub-paragraphs (i)–(vi) was amended in 1981 to indicate that
the list is not exhaustive. Vessels engaged in other kinds of operation which restrict
their ability to manœuvre as required by these Rules may be considered to be in this
category. A vessel transferring spare parts necessary for repairs whilst underway is
intended to be included in the category mentioned in (g)(iii) as it could be considered
to be transferring provisions.

The term 'minesweeping' in sub-paragraph (v) was changed to 'mine clearance'
by the 1981 amendments to include other operations such as minehunting.

(h) In 1968 IMO recommended (Resolution A 162 IV) that 'deep draught vessels' in
open waters, using channels which they would be unable to leave without risk of
grounding, should show the signals which are prescribed in Rule 28, so that they
could be recognised by other vessels.

At the 1972 Conference it was decided to include provisions relating to the so-called
deep draught vessels in the Rules. In considering how to define this type of vessel it
became apparent that it would not be satisfactory to specify a minimum size or
draught and it was decided not to restrict this category to very large ships. The more
appropriate term 'vessel constrained by her draught' was therefore used.

The main factor which must be taken into account is the space available for manœuvre rather than the depth of water beneath the keel, but the restriction of space must be
due to relatively shallow water which would not necessarily be a danger to other vessels
in the vicinity. The signals are mainly intended for use by vessels unable to make an
appreciable alteration of course, especially to starboard, due to passing between shoals
which are, however, deep enough not to restrict other vessels. The signals should only be
shown when the ability to alter course is severely restricted. On passing clear of the area
of relatively shallow water the signals must no longer be displayed.

A very large fully loaded vessel will not be justified in displaying the signals even
in crowded waters, or a traffic separation area, if there is sufficient deep water on
either side to permit course alterations.

In order to clarify this matter further the following item of guidance for the
uniform application of certain rules has been approved by the IMO Maritime Safety
Committee:

Clarification of the definition 'Vessel constrained by her draught', Rule 3(h). 'Not
only the depth of water but also the available navigable water width should be used as
a factor to determine whether a vessel may be regarded as constrained by her draught.
When determining this, due account should also be taken of the effect of a small
underkeel clearance on the manœuvrability of the vessel and thus her ability to
deviate from the course she is following. A vessel navigating in an area with a small

underkeel clearance but with adequate space to take avoiding action should not be regarded as a vessel constrained by her draught.'

In 1987 the fifteenth Assembly of IMO adopted an amendment to Rule 3(h) changing the words 'available depth of navigable water' to 'available depth and width of navigable water'. The purpose of this amendment was to confirm and strengthen the above clarification.

Special rules may apply in some harbours, rivers or inland water areas requiring signals to be shown by vessels over a certain size or exceeding a certain draught. In such cases the condition that the ability to deviate from the course must be severely restricted to justify showing the signals is unlikely to apply.

(i) The term 'under way' is sometimes used in a restricted sense as applying to a vessel which is actually moving through the water but this is not the meaning used in the Rules. Rule 35(a) prescribes sound signals for a power-driven vessel making way and Rule 35(b) specifies a different signal for a vessel under way but stopped and making no way through the water.

A vessel which is lying stopped is, nevertheless, expected to comply with the Rules of Sections II and III of the Steering and Sailing Rules. The following guidance to clarify this point has been approved by the IMO Maritime Safety Committee:

Clarification of the application of the word 'underway', Rule 3(i). 'When applying the definition of the term "underway" mariners should also have regard to Rule 35(b) where it is indicated that a vessel may be underway but stopped and making no way through the water.'

A vessel is only considered to be at anchor when the anchor is down and is holding. Vessels using an anchor to turn in the river, or riding to their chains with anchors unshackled, or dredging with the tide, or dragging their anchors have been held to be under way.

(k) In Rule 11 it is stated that the Rules in Section II of the Steering and Sailing Rules (Rules 12–18) apply to vessels in sight of one another.

They do not apply to vessels which have detected one another by radar but are not in visual sight.

(l) Sandstorms are included in the list of different conditions restricting visibility. Examples of 'other similar causes' are smoke from any vessel, afloat or ashore, including your own, and dust storms.

(m) The definition of 'wing-in-ground (WIG) craft' was added as an amendment to Rule 3 by the 22nd IMO Assembly in 2001. WIG craft are not to be considered as seaplanes or non-displacement craft. For the purpose of WIG craft there is a new provision included in Rule 18(f), (see page 87).

WIG craft look like aircraft, but are not aircraft and therefore the definition of 'seaplane' is not applicable to WIG craft. WIG craft are classified as dynamical support craft. The weight of a WIG craft in operation is mainly supported by a dynamic air cushion, which by engine thrust is created between the lower surface of an air foil and the water surface.

WIG craft have the capability of increasing the altitude of flight by enlarging the engine thrust, thus performing a jump to overcome and overfly obstacles on the earth surface. However increasing the altitude of flight of a WIG craft to perform a jump will significantly enlarge its fuel consumption and will therefore result in loss of economy.

Part B – Steering and sailing rules

Section I – Conduct of Vessels in any Condition of Visibility

RULE 4

Application

Rules in this Section apply in any condition of visibility.

COMMENT:

Part B – Steering and Sailing Rules deals with conduct of vessels. Rules 4 to 10, in Section 1 of Part B, are of a general nature and apply in any condition of visibility.

RULE 5

Look-out

Every vessel shall at all times maintain a proper look-out by sight and hearing as well as by all available means appropriate in the prevailing circumstances and conditions so as to make a full appraisal of the situation and of the risk of collision.

COMMENT:

Look-out man

On all but the smallest vessels a seaman should normally be posted on look-out duty from dusk to dawn and sometimes by day, especially when the visibility is restricted. Maintaining a proper look-out is an important element of safe watchkeeping. Requirements for safe watchkeeping are laid down in Chapter VIII of the International Convention on Standards of Training, Certification and Watchkeeping for Seafarers, 1978 as amended. (See pages 157–158.) Mandatory standards regarding watchkeeping, including standards for keeping a proper look-out, are contained in Part A, Chapter VIII of the STCW Code. (See pages 157–164.) Guidance on watchkeeping arrangements and principles to be observed is given in Part B of the STCW Code. (See pages 167–168.)

Before radar came into general use it used to be stressed in the Courts that the look-out man should be stationed forward, unless weather conditions made it impossible.

A Guide to the Collision Avoidance Rules. DOI: 10.1016/B978-0-08-097170-4.00002-7

This may still apply for a vessel without operational radar. It has been pointed out that a look-out stationed forward would have his attention distracted by conversations and activities of personnel on the bridge and would be more likely to hear fog signals coming from ahead. However, other factors such as the need to have a seaman immediately available in case of sudden emergency and the value of being able to communicate directly with the look-out man should also be taken into account.

Dea Mazzella–Estoril

I thought it right to ask the Elder Brethren who are advising me in this case what is their view of the practice of stationing the look-out man on the navigating bridge. They tell me that the look-out should certainly be stationed somewhere else in the ship; forward, if possible, if the weather conditions allow it. If, however, the weather is such as to forbid the possibility of a look-out being posted forward, then at least he ought to be stationed on the upper bridge. (Mr Justice Willmer, 1958)

Cabo Santo Tome–Cometa

She has definitely the noisier kind of engines, as is shown by the evidence of an independent ship as well as by one's own knowledge of diesel engines at full speed, and it seems to me, and I am advised by my assessor, that it was wrong in the circumstances of fog not to have a man on the look-out forward. She could by those means have obtained information of the approaching ship 150 feet farther forward in a fog of varying density. I think that was quite wrong on her part, and no doubt also the faintness of the whistle that she heard is to be attributed to the fact that she had not got a man properly placed. (Mr Justice Langton, 1933)

The Courts are likely to take into account the number of seamen available in addition to the state of visibility, probability of meeting other vessels and other factors when considering the sufficiency of look-out. No definite rules apply. However, even relatively small vessels may be expected to have a man posted on look-out duty at night in busy traffic lanes, or during periods of restricted visibility.

City of Naples

The sufficiency of look-out is a practical matter, and I am not satisfied that there was any necessity–still less do I think it the province of this House to lay down a hard and fast rule to that effect–for a special man on the forecastle to act as a look-out. (Lord Dunedin, 1921)

Spirality–Thyra

I do not think it is necessary for me to enlarge much more the reasons why this situation came about. I am satisfied that it came about because a bad look-out was being kept on board the Spirality. I appreciate the fact that she is only a small vessel which carries a crew of no more than eight hands. It is dangerous to lay down any absolute rules. All I desire to say – and I desire to say it with all the emphasis at my command, supported, as I am in this respect by the advice which I have received from the Elder Brethren – is that in no circumstances can it be right for a vessel of the class of the Spirality *to be left with only one man on deck–a man who had to do everything, control the ship, keep a look-out, and so forth–for a period which must, as I have said, have extended for the best part of ten minutes. I am advised by the Elder Brethren that it would be very difficult for a man in that position, having to keep his eyes on the compass, to keep the diligent look-out which is required, and required above all places in the River Thames. (Mr Justice Willmer, 1954)*

In the case of *Saxon Queen–Monmouthbrook* (1954) it was held that a small vessel with a crew of eleven men should have had a lookout on the forecastle head. The vessel was navigating without radar off the north east coast of England in visibility of about 400 metres. The master, officer of the watch and helmsman were inside the wheel-house and there was no seaman posted on look-out duty on the bridge.

Ocean look-out

There is some justification for relaxing the degree of look-out in the open ocean where other vessels are infrequently seen and are unlikely to be encountered so as to involve risk of collision. However, collisions occasionally occur in such areas, indicating the need for a proper look-out at all times. In October 1970, two vessels, each of approximately 10,000 tons gross, collided just before noon, in the middle of the Atlantic Ocean (approximate position 4°N 28°W). One vessel was on a voyage from India to eastern Canada and the other was bound for Spain from Brazil.

Duty of look-out

The look-out should report any lights, vessels or large floating objects which he sees, and, in low visibility, any fog signals which he hears. However, in crowded waters, he could not be expected to report everything he sees; he must use his discretion and report the lights or objects which are likely to bring risk of collision, especially small craft which may not have been observed from the bridge.

Shakkeborg–Wimbledon

You cannot report every light you see in the River Thames. You have to watch until you see a light, which, perhaps, you have seen before, becoming material, because if you are going to report every light in Gravesend Reach when coming up the River Thames the confusion would be something appalling to those in charge of the navigation; but you have to have a lookout to report every material light as soon as it becomes material. (Mr Justice Bargrave Dean, 1911)

All available means appropriate

The term 'proper look-out' has always been interpreted by the Courts as including the effective use of available instruments and equipment, in addition to the use of both sight and hearing. The use of binoculars, information received by VHF and automatic identification system (AIS) from a VTS station, shore radar station or from other ships would be included among "all available means appropriate".

Gorm–Santa Alicia

If the visibility was deceptive, as the pilot would have me believe, and he had not seen the North Sturbridge Buoy light, I find it difficult to understand why he did not resort to binoculars, or some other optical aid, to assist him. It is difficult, in my view, in any event, to understand why he did not use binoculars on seeing the approaching Gorm. *Apparently he remained behind closed windows in the wheelhouse. (Mr Justice Hewson, 1961)*

Bovenkerk–Antonio Carlos

I find that the Antonio Carlos *was at fault for bad look-out in the broadest sense; namely, faulty appreciation of VHF–information and total absence of radar look-out. (Mr Justice Brandon, 1973)*

Vechtstroom–Claughton

The question of the use that should properly be made of facilities that are provided is a matter I have discussed with the Elder Brethren, and, if I may say so, I am in wholehearted agreement with them that these facilities of radar advice are made and supplied and established for the greater safety of shipping in general and for greater accuracy in navigation; in fact, this particular Seacombe radar station was established by one of the ferry–operating corporations itself. We can only presume that it was put there for a good purpose and to be used in such conditions as prevailed on that morning. A vessel which deliberately disregards such an aid when available is exposing not only herself, but other shipping to undue risks, that is, risks which with seamanlike prudence could, and should, be eliminated. As I see it, there is a duty upon shipping to use such aids when readily available–and when I say 'readily available' I am not saying instantly available–and if they elect to disregard such aids they do so at their own risk. (Mr Justice Hewson, 1964)

Radar not working properly

There should be no obligation to use radar in restricted visibility if the set is not functioning properly, provided it can be shown that there was a genuine fault. Everything possible should be done to have the set repaired and brought back into use.

In an American case *Pocahontas Steamship Company–Esso Aruba*, 1950, the judge said:

There might well be times when the continued use of radar by a navigator who was uncertain of the results he was observing and unwilling to place reliance thereon might well be foolhardy and hazardous.

The radar may have to be temporarily disregarded due to such things as excessive interference, or even switched off if its continuing use may damage the set.

However, the following comment was made in the US Appeal Court with reference to the above quotation from the judgment on the *Esso Aruba:*

This does not mean that, in the face of the fact that a properly functioning radar will give useful and necessary information, the master had a discretion to decide that it will not give such information and turn off his radar. A master has no more discretion to disregard this aid to navigation than he has to disregard the use of charts, current tables and soundings where the circumstances require the use thereof.

If a vessel carries properly functioning radar equipment and she is in or approaching an area of known poor visibility, there is an affirmative duty to use the radar. (Judge Medina, 1959)

Use of radar in clear visibility

In American Courts vessels colliding with oil drilling platforms have been held to be at fault for not using radar at night in clear visibility when passing through areas where there were known to be numerous structures which are not always adequately lit. The radar should, preferably, be kept in use for the purpose of keeping a general lookout in coastal waters, and other areas where regular traffic is likely to be encountered, especially at night. Rule 6(b)(vi) refers to the use of radar for assessing visibility (see page 17).

Visual look-out still necessary

The use of radar does not dispense with the need for a good visual look-out.

Anneliese-Arietta

One of the matters which will have to be considered is the effect, if any, on this collision of the Arietta *relying on observation with her relative motion radar without having apparently any visual look-out at all. That is clearly an important matter of seamanship, on which we have thought it right to consult our assessors. The question put to them on this occasion was: 'Was it seamanlike for the* Arietta *to rely on relative motion radar observation only and to have no visual look-out?' and the answer was: 'No.' For myself, I accept that answer without the least hesitation. The use of radar is by no means to be despised, especially in fog, where it has been described as an extra eye; but the human eye can sometimes see more quickly than radar even in fog, and so is able to appreciate the position in less time than is needed to examine the recordings of a radar. This was a relative motion radar, and to get a true picture from such a radar plotting was required, and plotting takes time. I repeat that I accept without any hesitation the advice we have received about this that the* Arietta *should have had a good visual look-out in addition to the radar. (Lord Justice Karminski, 1970)*

Full appraisal of the situation

In order to keep a proper look-out the officer of the watch, or person in charge, must also pay attention to what is happening on his own vessel keeping a check on the steering and seeing that equipment required for keeping the vessel on course is functioning correctly.

Staffordshire–Dunera

Where, in my judgment, she was at fault, was in having a very bad look-out, and a bad look-out in every possible sense of the term. It seems to me that it comes within the term 'bad look-out' when I say that she was at fault for failing to take proper precautions to meet the situation in the event of the compass breaking down again, as it in fact did. It was, in my judgment, bad look-out on the part of this young third officer in failing to appreciate, long before he did appreciate it, what was happening, namely, that his vessel was falling off to starboard, and in failing to appreciate what the probable cause of the falling off was. It was bad look-out on the part of the quartermaster, when he knew perfectly well that the compass had stuck again, not to report the matter at once to the officer in charge. It was bad look-out on the part of the officer to take no steps himself, whether by going to the standard compass or otherwise, to check up on what was happening and what was the course of his vessel. (Mr Justice Willmer, 1948)

Several collisions have occurred as a result of a failure of steering gear, automatic pilot or gyro compass. In September 1964, the British cargo ship *Trentbank* developed a fault in the automatic pilot as she was overtaking the Portuguese tanker *Fogo* in the Mediterranean. The *Trentbank* swung across the bow of the *Fogo*. The following comment was made in the judgement with reference to the look-out:

I ought not to leave this part of the case without observing how lamentable was the attitude of the master of the Trentbank *and her chief officer towards the system of automatic steering. The master had given no orders to ensure that somebody was on look-out all the time. The chief officer, according to his own story, saw nothing wrong in undertaking a clerical task and giving only an occasional glance forward when he knew that there was other shipping about and that he was the only man on board this ship who was keeping any semblance of a look-out at all. Automatic steering is a most valuable invention if properly used. It can lead to disaster when it is left to look after itself while vigilance is relaxed. It is on men that safety at sea depends and they cannot make a greater mistake than to suppose that machines can do all their work for them. (Mr Justice Cairns, 1967)*

Anchor watch

The duty to keep a proper look-out applies also when a vessel is at anchor, especially if there is a strong tide running, or if other vessels are likely to be passing by.

Gerda Toft–Elizabeth Mary

It may be that a seaman cannot help his anchor dragging in certain circumstances, but what he can do, and what he has a duty to do, is to keep a good look-out and take prompt measures to stop the dragging if and when it does occur. The failure of the

Gerda Toft *to take timely measures in this case was due, as I find, to bad look-out. As I have already said, both her officers were in the chartroom at the material time, and the only look-out was that of the extremely ineffective seaman, who remained on deck, and to whom I have already referred. Because of this bad look-out those in charge of the* Gerda Toft *as it seems to me, had no real idea of what was happening, and, therefore, failing to appreciate the situation, failed to take any adequate steps to arrest the dragging of their vessel. (Mr Justice Willmer, 1953)*

RULE 6

Safe speed

Every vessel shall at all times proceed at a safe speed so that she can take proper and effective action to avoid collision and be stopped within a distance appropriate to the prevailing circumstances and conditions.

In determining a safe speed the following factors shall be among those taken into account:

(a) **By all vessels:**
  (i)   **the state of visibility;**
  (ii) **the traffic density including concentrations of fishing vessels or any other vessels;**
  (iii) **the manœuvrability of the vessel with special reference to stopping distance and turning ability in the prevailing conditions;**
  (iv) **at night the presence of background light such as from shore lights or from back scatter of her own lights;**
  (v)   **the state of wind, sea and current, and the proximity of navigational hazards;**
  (vi) **the draught in relation to the available depth of water.**
(b) **Additionally, by vessels with operational radar:**
  (i)   **the characteristics, efficiency and limitations of the radar equipment;**
  (ii) **any constraints imposed by the radar range scale in use;**
  (iii) **the effect on radar detection of the sea state, weather and other sources of interference;**
  (iv) **the possibility that small vessels, ice and other floating objects, may not be detected by radar at an adequate range;**
  (v)   **the number, location and movement of vessels detected by radar;**
  (vi) **the more exact assessment of the visibility that may be possible when radar is used to determine the range of vessels or other objects in the vicinity.**

COMMENT:

The wording of Rule 6 and its location with respect to the other Rules should leave no doubt that the setting of a safe speed is a prerequisite in all conditions of visibility. It is, of course, in restricted visibility that the need to moderate the speed generally applies and the state of visibility is listed first among the factors to be taken into account in determining a safe speed. Unlimited visibility should not, however, be considered as justifying full speed under all circumstances.

Every vessel

The requirement to proceed at a safe speed at all times applies to every vessel. This point may have special significance with respect to vessels constrained by their draught, and to some vessels restricted in their ability to manœuvre, which may not be justified in maintaining a high speed when other vessels are in the vicinity because of their limited manœuvrability.

Safe speed

The term 'safe speed' has not been used in previous regulations. It replaces the term 'moderate speed' which was only related to the conditions of restricted visibility. A new term was necessary which would be applicable at all times and which would not preclude the setting of a high speed in appropriate circumstances.

The word 'safe' is intended to be used in a relative sense. Every vessel is required to proceed at a speed which could reasonably be considered safe in the particular circumstances. If a ship is involved in a collision it does not necessarily follow that she was initially proceeding at an unsafe speed. In clear visibility collisions can generally be attributed to bad look-out, or to wrongful action subsequent to detection, rather than to a high initial speed.

At all times

In order to maintain a safe speed at all times a continuous appraisal of changes in circumstances and conditions should be made and any necessary alterations of speed must be instantly put into effect. It is important that watchkeeping officers should not be obliged to communicate with the master before using the telegraph as the resulting delay could have serious consequences. The IMO Recommendation relating to watchkeeping states that 'the officer of the watch should bear in mind that the engines are at his disposal and he should not hesitate to use them in case of need. However, timely notice of intended variations of engine speed should be given when possible'.

A relatively high speed might be accepted as being initially safe for a vessel using radar in restricted visibility in open waters provided prompt action is taken to bring the speed down when radar information shows this to be necessary.

Kurt Alt–Petrel

While, if properly used and can be relied upon to indicate all potential dangers in ample time safely to avoid them, it may give some justification for a speed in restricted visibility which would otherwise be immoderate, such a speed can only be justified so long as it is safe to proceed and provided that timely action is taken to reduce it or take off all the way in the light of the information supplied or to be inferred from the radar. (Mr Justice Hewson, 1962)

Proper and effective action

A vessel may be unable to take proper and effective action due to the speed being too high or, in some circumstances, too low. For instance, in restricted visibility the speed of a vessel without operational radar may be too high to enable effective avoiding action to be taken on sighting another ship or, in the case of a vessel using radar, too high to enable proper assessment to be made after detection especially after the detection of small vessels. On the other hand, in certain circumstances, it may be dangerous to reduce speed to such an extent that the steering becomes ineffective.

In the *Ring–Orlik*, 1964, the *Ring* was found at fault for losing steerage-way and falling off her course, in the Elbe, when another vessel was close astern and overtaking.

But, in my view, it was the duty of those on the bridge of the Ring to appreciate that they had lost steerage-way and were going off course and it was their duty to correct it by appropriate engine and helm movement. The Ring's failure in those respects was a cause of the collision. (Sir Jocelyn Simon, President of the Court)

Within a distance appropriate to...

The term 'moderate speed' was previously interpreted as meaning a speed which would enable a vessel to be stopped within half the range of visibility.

Glorious–Florida

There is an excellent rule that we sometimes come across in motor collision cases and which we act on – that if there is a difficulty in seeing you ought to be ready to stop within the limits of visibility; and obviously a boat that goes on fast in dense fog will not be able to stop within the limits of visibility. That the other obstacle, so to speak, is not – as is often the case in motor cases – a fixed barrier which does not move, but is also a ship which is likewise moving, must cut down that limit of visibility by one-half, and each boat should be able to stop well within its limits of visibility. (Lord Justice Scrutton, 1933)

Umbria (United States Supreme Court)

The general consensus of opinion in this country is to the effect that a steamer is bound to use only such precautions as will enable her to stop in time to avoid a collision, after the approaching vessel comes in sight, provided such approaching vessel is herself going at the moderate speed required by law. (Mr Justice Brown, 1897)

However, it has since been held in the British Courts that this is not a rule of law (Morris v. Luton Corporation, 1946); each case must be judged with regard to the existing circumstances and conditions. The rule might be appropriate for a vessel without radar in areas where small craft are likely to be encountered but a ship which is making proper use of radar in the open ocean is not expected to take all way off when the fog becomes so dense that it is not possible to see beyond the forecastle head. Half the range of visibility might even be too large to be an appropriate stopping distance if the visibility is approximately 1 mile, especially for a vessel without radar, as this would mean that good stopping power could justify speeds in excess of 20 knots. Other factors, apart from visibility and stopping ability, must be taken into account. High speed will give little opportunity for assessing what action should be taken when a vessel is sighted or detected at short range.

In a case which came before the High Court of Justice in London in 1972 (*Hagen– Boulgaria*) the Elder Brethren were asked to advise what would be a proper speed for a cargo ship, 135 metres in length capable of 17 knots, at night, without radar, in the English Channel where much traffic could be expected, in visibility which for some time had been about 1 mile. They replied that it would be about 6 to 7 knots. The Elder Brethren were also asked what would be a proper speed for the other vessel which was using radar in visibility of about 6 cables and they replied that it would be about 8 to 9 knots. The ship, 108 metres in length, had diesel engines which gave a speed of 13 ½ knots. Their advice, in the second instance, was qualified by saying that even if the vessel concerned had been going at that speed a further reduction should have been made on running into thick fog and seeing a close quarters situation developing. In each case the Judge accepted the advice.

The above example has only been included to give some indication of how the Courts might interpret the term 'safe speed' for different vessels. Too much importance should not be attached to the specific values quoted as so much depends upon the circumstances which apply in each particular case.

Attempts have been made to quantify speed in relation to the range of visibility and other factors but discussions at the IMO meetings did not result in an acceptable method of determining what value of speed would be appropriate to the conditions. The list of factors to be considered when determining a safe speed is intended to assist the mariner by drawing attention to points which might otherwise have been overlooked. The factors are not meant to be in order of importance and the list is not exhaustive.

FACTORS TO BE TAKEN INTO ACCOUNT

By all vessels

Most of the factors are generally self-evident. The state of visibility is obviously of major importance. Rule 19 requires a power-driven vessel to have her engines ready for immediate manœuvre in restricted visibility and every vessel, when risk of collision exists, to reduce her speed to the minimum at which she can be kept on her course when a fog signal is heard forward of the beam, or when a close quarters situation cannot be avoided with another vessel forward of the beam. This Rule therefore places further limitations on the value of safe speed in restricted visibility (see pages 93–95).

Information concerning stopping distances and turning circles is now supplied to many vessels and navigating officers are expected to be familiar with the manœuvring characteristics of their own ship. The distance that a vessel will cover in a crash stop before being brought to rest from full speed is likely to be between 5 and 15 ship lengths, depending upon speed, displacement, type of machinery, etc. Some general guidance on manoeuvring characteristics is given on pages 177–179.

The manoeuvrability is to be taken into account with reference to the prevailing conditions. A vessel which is restricted in her ability to manoeuvre because of the nature of her work may not be justified in going at a high speed in regions of high traffic density and when approaching a relatively slow vessel so as to involve risk of collision.

The reference to draught is intended to cover the possible restriction of manoeuvring space due to shallow water in the vicinity, or the hydro–dynamic effects, such as bow cushion, bank suction and interaction between ships, which can generally be eliminated or reduced by a reduction of speed.

Additionally, by vessels with operational radar

From the context of the Rule it is apparent that the term 'operational radar' means radar in use. However, it must be appreciated that radar is required to be used, when appropriate, both for keeping a proper look-out and for determining risk of collision (see Rules 5 and 7) provided, of course, that it is in working order.

In restricted visibility a vessel making proper use of radar will normally be justified in going at a higher speed than that which would be acceptable for a vessel which does not have the equipment but not usually at the speed which would be considered safe for good visibility (see the example on page 20). Rule 6(b) requires several factors to be taken into account. Some of the factors were included in the Annex to the 1960 Regulations which merely gave recommendations on the use of radar information as an aid to the avoidance of collision.

Characteristics, efficiency and limitations

A considerable choice of radar equipment is available ranging from the relatively small, low cost, installations which are intended for use on small vessels to the highly sophisticated computer-aided systems fitted to some of the larger ships. Even the most efficient equipment cannot be regarded as a complete substitute for the human eye. Radar may fail to detect small targets, alterations of course made by other vessels are usually less apparent and the use of radar bearings is more likely to result in a faulty appreciation of risk of collision than visual bearings taken by compass. However, the use of radar in clear visibility does give some advantages such as range indication and, with some systems, the prediction of the distance of nearest approach and an indication of the effectiveness of proposed manoeuvres to avoid collision.

All vessels of 10,000 gross tons and upwards are required to be fitted with automatic radar plotting aids (ARPA) by the IMO Safety of Life at Sea (SOLAS) Convention. Such equipment is being fitted to an increasing proportion of smaller ships. When advanced equipment of this type is provided it is expected to be put to effective use.

The efficiency of the equipment for the purpose of detecting the presence of other vessels and determining whether risk of collision exists must also be related to the competence of those observing it and the way it is being used. Occasional glances at the radar screen would hardly constitute proper use of the equipment to justify a high speed in restricted visibility.

Norefoss–Fina Canada

When reliance is placed on the radar, it cannot be too strongly emphasized that a continuous radar watch should be kept by one person experienced in its use, as this officer was. (Mr Justice Hewson, 1962)

Niceto de Larrinaga–Sitala

High speeds at collision cause much greater damage than low speeds. High speeds before collision give less time to appreciate properly the development of the situation. Therefore, if radar is relied upon it must be properly used. If you rely upon the extended and accurate look-out which is provided by radar to justify immoderate speed, you must be careful to see that you use your radar properly and with seamanlike prudence upon the indications and inferences which are given by it, or may be drawn from the data supplied by it. (Mr Justice Hewson, 1963)

The radar should be properly set up, making such adjustments to the controls as may be necessary to achieve maximum efficiency. To check that this efficiency is being maintained the performance monitor should be used at frequent intervals.

If any shadow sectors or blind arcs are suspected, or known, to be present, the vessel should be swung off course for a short period at regular intervals so that they may be examined. The US Coast Guard investigation of the collision between the vessels *Sparrows Point* and *Manx Fisher* found that the *Manx Fisher* had approached within the shadow sector of the radar of the *Sparrows Point.*

Range scale

Constraints may be imposed by every range scale that can be used. When using the longer range scales definition and discrimination are reduced and small targets are less likely to be detected, whereas shorter range scales do not permit early detection of targets and do not enable the observer to obtain an overall assessment when several vessels are in the vicinity. The range scale which is most suitable for the locality should be selected but the scale should be changed at regular intervals. The scale should not be changed when there is a dangerous target at close range.

When two radar displays are available and in use it may be advantageous to select a different range scale on each display to avoid the necessity of switching scales.

Nassau–Brott

If the master of the Nassau *was relying upon radar to justify his speed in reduced visibility it was not good seamanship to have kept his radar permanently on the short range. It is a matter which I have thought about and discussed with the Elder*

Brethren, and we are agreed upon what I am about to say. They should have extended the range periodically at intervals appropriate to the circumstances to inform themselves of the general situation and, in particular, of the probable effect of the approach of otherwise invisible vessels upon the action of the vessel known to be, and seen to be, ahead of them, the Brott. *(Mr Justice Hewson, 1963)*

Interference

Proper use of controls should normally enable ship echoes to be distinguished from clutter due to waves and precipitation but such interference may sometimes be so severe that even large targets may be obscured. Echoes from small craft are especially likely to remain undetected when such interference is present.

The effect of rain clutter is much less when using 10 cm wavelength than when using 3 cm. This is often also the case with clutter caused by sea return. Vessels fitted with two radars, one of each wavelength, would be expected to make use of the 10 cm wavelength for detecting other vessels in conditions likely to cause severe clutter, particularly in heavy tropical rain squalls.

In 1979 the *Atlantic Empress* and the *Aegean Captain*, two very large laden tankers, collided off Tobago near the edge of a tropical rain squall. The *Aegean Captain* had just passed through the heavy rain. Both vessels were proceeding at full speed and in each case detection was not achieved until the range was less than 2 miles. At the subsequent inquiry in Greece it was considered that both vessels did not make effective use of their radar equipment and were proceeding at excessive speed for the conditions of visibility.

Small craft and ice

Minor targets such as small coastal vessels and trawlers should normally be detected at distances greater than 6 miles, provided the set is properly adjusted, but yachts, open boats and other small craft, especially boats of fibreglass construction, usually give poor echoes and may not be detected in time to take effective avoiding action. The fitting of an efficient radar reflector is likely to considerably increase the probability of being detected and may double the range of detection for a small vessel.

The Safety of Life at Sea Convention (SOLAS) requires all ships of less than 150 gross tons to be fitted, if practicable, with a radar reflector or other means, to enable detection by radar at both 9 and 3 GHz. The reflector should be of an approved type complying with minimum performance standards, preferably mounted at a minimum height of 4 m above water level.

Following the loss of the yacht *Ouzo* due to a collision off the Isle of Wight in August 2006 the Marine Accident Investigation Branch commissioned a study into marine radar reflectors. The report of the study included a recommendation that yachtsmen should always fit a radar reflector that offers the largest radar cross section practicable for the vessel with a minimum radar cross section of $2\,\text{m}^2$. It is recommended that poorly performing radar reflectors should not be fitted as it is possible that the user could be lulled into a sense of false security believing the chance of detection has been enhanced. The United Kingdom Government subsequently issued a Marine Guidance Note (MGN 349) which recommends that yachtsmen should permanently fit, not just carry on board, a radar reflector or radar target enhancer that offers the largest radar cross section practicable for their vessel. It is emphasised that the reflector must be mounted at a minimum height of 3 m (preferably 4 m) to take it out of any wave obscuration effects and give a potential detection range of 5 nm.

The following two collisions, which occurred off the coasts of the United States, illustrate the danger of placing too much reliance on radar in areas where small craft may be encountered. Both of the fishing vessels referred to sank with resulting loss of life.

In September 1959, the *SS Mormacpine* was approaching the Straits of Juan de Fuca in visibility estimated at between 500 and 1,000 yards. Full speed, of approximately 11 knots, was being maintained, but the engines were on stand-by. The radar was on the 8 mile range and appeared to be working satisfactorily. When a fog signal was heard ahead the engines were immediately stopped. The master checked the radar and found no targets. Approximately 1½ minutes later the look-out reported sighting a vessel 1,000 feet ahead, fine on the starboard bow. This later proved to be the *Jane*, a 49 foot, wood hull fishing vessel. Although the engines of the *Mormacpine* were immediately put full astern, this was not sufficient to prevent collision.

In April 1961, the *South African Pioneer* was on a voyage from Charleston, South Carolina to New York. In visibility of approximately 1½ miles the engines were placed on stand-by, resulting in a speed of approximately 10 knots. The radar was on the 8 mile range, and sea clutter was observed to extend approximately 3 miles out from the centre of the display. No targets had been observed by radar, but a light was sighted 10° on the starboard bow. This later proved to have been the sidelight of the *Powhatan*, a 78 foot wood hull fishing vessel. Despite drastic helm and engine action the *South African Pioneer* was unable to avoid collision.

Tests carried out by the US Coast Guard have shown that small icebergs of sufficient size to be dangerous to navigation should normally be detected at a range of about 4 miles. If the part above surface is particularly smooth, however, they may remain undetected, especially when appreciable sea clutter is present. In regions where small craft and ice are likely to be encountered the speed should be low enough to enable the vessel to be stopped well within the range of visibility.

Number, location and movement of vessels detected

In determining a safe speed the mariner must take account of the traffic situation in his vicinity. The greater the number of targets indicated on the radar display the more difficult it may be to determine risk of collision and to assess the effect of possible manœuvres, although some radar systems are capable of providing information of this kind. Vessels detected ahead, or fine on the bow, proceeding in the opposite direction, will obviously present a greater threat than vessels observed to be approaching from abaft the beam with a low closing speed.

Where traffic separation schemes apply a ship using a traffic lane in restricted visibility may find it dangerous to reduce to a very low speed when proceeding in the general direction of traffic flow as this may result in her being frequently overtaken by other ships passing at close distances. However, a high speed is not necessarily justified in order to keep pace with other vessels in a traffic lane. The possibility of encountering crossing vessels and small craft must be taken into account.

Assessment of visibility

When fog or mist is considered likely to develop the radar should be in operation. It may be possible to determine the extent of the visibility by observing the radar ranges at which other vessels or navigation marks are first visually sighted, or at which they disappear from view. At night the probable presence of fog may be indicated by failure to see the lights of a vessel which gives a strong echo within the normal visual range.

In areas such as the North Western Atlantic, North Pacific and North Western Europe where there is a high incidence of fog particular caution is necessary, but during the hours of darkness watchkeeping officers should always be mindful of the possibility of the visibility being restricted, even in areas where fog occurs infrequently, and should use the radar for the purpose of determining visibility whenever this seems to be necessary.

In the early morning of the 16th October, 1965, the tanker *Almizar* was proceeding towards the Persian Gulf at full speed on a northerly course off the coast of Oman. The sea was calm and the visibility had previously been excellent. On the radar display the second officer observed the echo of another vessel right ahead and assumed that it was an unlighted dhow as no lights were sighted. When the range closed to three miles he changed to manual steering and ordered the helmsman to alter course 40° to starboard. He subsequently realised that there was fog and rang standby on the telegraph. The echo was in fact caused by an approaching tanker, the *John C. Pappas*, of 237 metres length. The two vessels collided causing serious damage.

When the case came to the High Court in London both ships were found to be at fault in several respects. The *Almizar* was blamed for entering the fog at too high a speed and reducing too slowly, also for keeping a poor radar lookout in mistaking a large ship on an opposing course for a dhow.

RULE 7

Risk of collision

(a) **Every vessel shall use all available means appropriate to the prevailing circumstances and conditions to determine if risk of collision exists. If there is any doubt such risk shall be deemed to exist.**

(b) **Proper use shall be made of radar equipment if fitted and operational, including long-range scanning to obtain early warning of risk of collision and radar plotting or equivalent systematic observation of detected objects.**

(c) **Assumptions shall not be made on the basis of scanty information, especially scanty radar information.**

(d) **In determining if risk of collision exists the following considerations shall be among those taken into account:**
   (i) **such risk shall be deemed to exist if the compass bearing of an approaching vessel does not appreciably change;**
   (ii) **such risk may sometimes exist even when an appreciable bearing change is evident, particularly when approaching a very large vessel or a tow or when approaching a vessel at close range.**

COMMENT:

Risk of collision

Rules 12,14, 15 and 18 require one vessel to keep out of the way of another when risk of collision exists. When one of two vessels in sight of one another is required to keep out of the way the other must keep her course and speed (Rule 17). The question arises as to how far apart the vessels must be before risk of collision should be considered to exist and the obligation to keep course and speed first begins to apply to the privileged vessel.

The 1972 Conference rejected a proposed definition that 'risk of collision' exists between vessels when their projected courses and speeds place them at or near the same location simultaneously. Had this definition been accepted a vessel detecting another at long range, slowly approaching from the port side with little change of bearing, would have been obliged to keep her course and speed for a long period, possibly several hours.

In the Courts of the United Kingdom and other countries risk of collision has not been held to apply at long distances when there is a low speed of approach. As the above definition was not accepted the previous Court interpretation should also apply to the 1972 Rules.

Banshee–Kildare

Now at what period of time is it that the Regulations begin to apply to two ships? It cannot be said that they are applicable however far off the ships may be. Nobody could seriously contend that if two ships are six miles apart the Regulations for Preventing Collisions are applicable to them. They only apply at a time, when, if either of them does anything contrary to the Regulations, it will cause danger of collision. None of the Regulations apply unless that period of time has arrived. It follows that anything done before the time arrives at which the Regulations apply is immaterial, because anything done before that time cannot produce risk of collision within the meaning of the Regulations. (Lord Esher, 1887)

The above case was heard in the nineteenth century when ships were relatively slow, but this extract from the judgment sets out clearly when the Regulations begin to apply. The two vessels concerned were involved in an overtaking situation in Dublin Bay, their speeds being respectively 6 and 7 knots. The distance at which risk of collision begins to apply might well be considered to be greater than 6 miles between vessels approaching one another at high speeds, in the open sea, on reciprocal or nearly reciprocal courses. The distance must depend very much on circumstances and particularly on the speed of approach. In rivers and harbours where vessels frequently have to change course risk of collision may only be considered to apply at relatively short distances.

All available means

The requirement to use all available means appropriate is also included in Rule 5, but for a different purpose. In determining whether risk of collision exists with a vessel which has been visually sighted the taking of compass bearings may be especially important. In clear visibility in the open sea the use of radar and associated equipment is more likely to be considered necessary for determining risk of collision, with a vessel seen to be approaching, than for the purpose of keeping a general look-out. The radiotelephone may be used to advantage in certain circumstances for the purpose of clarifying a situation involving two vessels and indicating intentions, in addition to its use for determining information about the location and movement of other vessels as an aid to the keeping of a proper lookout. When using a traffic separation scheme, information received by VHF radiocommunication and/or automatic identification system (AIS) concerning the movement of other vessels, particularly about vessels moving against the established direction of traffic flow, may give early indication of impending risk of collision.

The radiotelephone has been proved to be of value in the Great Lakes and other areas where the number of collisions per year has shown a marked decrease since pilots started to use the equipment. In rivers, canals and inland waters it is possible to ensure that all ships are supplied with suitable radiotelephone equipment which can be used by pilots familiar with an agreed procedure and speaking the same language. Identification of other vessels is facilitated by reference to navigation marks and by communication with the shore radar station or the controlling authority.

In international waters radiotelephony is occasionally used to advantage, but the difficulties of identifying an approaching vessel from other vessels in the vicinity and of communicating with a vessel of different nationality place severe limitations on its use. The confusion which might occur in an attempt to communicate with an unidentified vessel, approaching rapidly with no appreciable change of compass bearing, could be a contributory cause of collision. These problems may eventually be overcome and there is little doubt that the use of the radiotelephone for the purpose of collision avoidance will be of increasing importance in the future.

The United Kingdom Government has issued a Marine Guidance Note (MGN 324) to draw the attention of mariners to the risks involved when VHF radio is used as a collision avoidance aid. The Notice stresses the problems of identification and communication and makes the point that valuable time may be wasted in attempting to make radio contact instead of concentrating on the assessment of collision risk and the need for action. Reference is also made to the further danger of proposing, by VHF radio, to take action which is not in compliance with the Collision Regulations.

Angelic Spirit–Y Mariner

I accept the evidence of the master and the third officer of Angelic Spirit *that the master attempted to contact* Y Mariner *by VHF. The third officer of* Y Mariner *said in his statement that he tried to contact* Angelic Spirit. *He may have made some effort to do so, but if he did, it was ineffective. I do not, however, think that either ship's efforts to contact the other by VHF affect liability for the collision. It has been emphasised many times that ships should be navigated by reference to the Collision Regulations and not by VHF. (Mr Justice Clarke, 1994)*

Vessels have often been criticised by the Courts in the United Kingdom and elsewhere for the misuse of VHF but it has also been accepted that VHF may be used to advantage for purposes of collision avoidance in some circumstances.

In the *Mineral Dampier–Hanjin Madras, 2001,* Lord Philips made the following comment with reference to previous criticism of the misuse of VHF made by Mr Justice Sheen *(Majola Il-John M)*:

But we do not think that Mr Justice Sheen's comments should be read as an embargo on all VHF communications about navigation between two vessels which are passing or are approaching a close quarters situation. The Admiralty Court tends to experience cases where VHF conversations have led to disastrous misunderstanding. It does not become aware of cases where an exchange of VHF information has assisted safe navigation. As the Judge observed in this case, in some circumstances VHF conversations can be useful in order to exchange information between vessels. It is, of course, important that before paying regard to information received from another vessel there should be no doubt as to which vessel is sending the information.

Where two vessels approaching one another are in VHF communication it can in some circumstances be helpful if the vessel which is required to give way informs the other vessel of action being taken in order to comply with the collision regulations. Equally there may be circumstances in which the stand-on vessel is justified in asking the give-way vessel what action the latter is taking in order to comply with the collision regulations. Where two vessels are approaching each other in restricted visibility in circumstances where r. 19 applies a vessel which is taking avoiding action in compliance with that rule may well assist the other vessel if it informs that vessel on the VHF of the action being taken.

Automatic identification systems (AIS) are now being fitted to vessels, in accordance with the carriage requirements of the Safety of Life at Sea Convention (SOLAS). Data to be provided automatically by AIS includes ship identification, ship type, and position, course and speed of vessel. As AIS can be used to advantage for collision avoidance, such as in determining the identity of another vessel and in more rapid detection of changes of heading, vessels may be expected to make use of the equipment in appropriate circumstances. However, it should be borne in mind when using AIS for collision avoidance that not all vessels are required to be fitted with AIS.

Resolution A.917(22), adopted by IMO in 2001, gives guidelines on the operational use of AIS. The Resolution includes the following guidance on the use of AIS in collision avoidance situations:

The potential of AIS as an anti-collision device is recognized and AIS may be recommended as such a device in due time.

Nevertheless AIS information may be used to assist in collision avoidance decision-making. When using the AIS in the ship-to-ship mode for anti- collision purposes, the following cautionary points should be borne in mind:

> *1. AIS is an additional source for navigational information. AIS does not replace, but supports, navigational systems such as radar target-tracking and VTS; and*
> *2. the use of AIS does not negate the responsibility of the OOW to comply, at all times, with the Collision Regulations.*

The user should not rely on AIS as the sole information system, making use of all safety-relevant information available.

The use of AIS on board ship is not intended to have any special impact on the composition of the navigational watch, which should continue to be determined in accordance with the STCW Convention.

Once a ship has been detected, AIS can assist in tracking it as a target. By monitoring the information broadcast by that target, its actions can also be monitored. Changes in heading and course are, for example, immediately apparent, and many of the problems common to tracking targets by radar, namely clutter, target swap as ships pass close by and target loss following a fast manœuvre, do not affect AIS. AIS can also assist in the identification of targets, by name or call sign and by ship type and navigational status.

UK Marine Guidance Note (MGN 324), gives operational guidance for AIS. Included in this guidance is a warning that faulty data input to AIS could lead to incorrect or misleading information being displayed on the status of other vessels.

Findings of MAIB on use of AIS:

A collision in the East China Sea between the *Hyundai Dominion* and the *Sky Hope* in June 2004 was the subject of an investigation by the Marine Accident Investigation Branch (MAIB) of the United Kingdom. One of the findings of the MAIB was that the officer of the watch on *Hyundai Dominium* wasted time in sending an AIS text message and in VHF discussions with the other ship rather than taking positive and early action.

A collision in the Taiwan Strait between *Lykes Voyager* and *Washington Senator* in April 2005 was also the subject of an MAIB Investigation. Both ships were fitted with AIS but it was considered that AIS information was not fully utilised. When course alterations were made by both vessels before the collision the manœuvres were not immediately apparent to the other vessel due to the lag in the ARPA system. Had either master monitored the AIS information the manœuvring of the other ship would have been quickly apparent.

Appropriate to the prevailing circumstances

The phrase 'appropriate to the prevailing circumstances and conditions' indicates that it is not always necessary to use radar to determine whether risk of collision exists.

However, paragraph (a) should not be interpreted as only requiring radar to be used in restricted visibility. Visual compass bearings are generally preferable to radar bearings when vessels are in sight of one another, but the use of radar enables ranges to be taken. A stand-on vessel can use radar to determine whether the vessel required to keep out of the way is not taking appropriate action or is so close that collision cannot be avoided by the action of the give-way vessel. A vessel which is obliged to pass close to another vessel when overtaking can use radar to ensure that there is a safe passing distance. The effects of interaction and the possibility of a steering defect should be taken into account.

Vessels in visual sight of one another have even been considered to be at fault for not using radar to determine whether risk of collision exists in cases to which the 1960 Rules were applicable. In the *Statue of Liberty–Andulo* (1970) the Assessors advised that both the give-way vessel and the stand-on vessel should have used radar in a fine crossing situation off the coast of Portugal. In the *Verdi–Pentelikon* (US Court, 1970) the stand-on vessel was blamed for failing to use radar in an area of dense traffic, such as the Straits of Gibraltar, to obtain information regarding a vessel which had been observed to be on a constant bearing on the port bow.

In regions of heavy traffic the need to use radar in good visibility would apply to all vessels, but it may apply especially to ships fitted with the more sophisticated radar systems which incorporate such additional facilities as the ability to indicate whether there is risk of collision with several targets and to determine the effectiveness of proposed manœuvres. In such areas radar can be used to assess the general traffic situation in all conditions of visibility.

Proper use of radar equipment

In the 1960 Regulations there was no direct reference to radar in any of the actual Rules but recommendations on the use of radar information as an aid to avoiding collision were given in an Annex. The importance of radar for the purpose of collision avoidance has now been recognised by incorporating the former recommendations into the Rules to make them requirements. Rules 6, 7, 8 and 19 contain specific references to the use of radar and there is an important implied reference in Rule 5.

Proper use of radar to obtain early warning of risk of collision requires that all controls should be at their optimum settings and that the appropriate range scale should be used. In addition the choice of display may be important. It has been generally recommended that a stabilised display should be used where possible. This arrangement has the advantage of enabling compass bearings to be read off directly, and the echoes do not become blurred due to an alteration of course.

Where a true motion display is provided it should be used in conditions favourable to this type of presentation. An alteration of course made by another vessel moving at fairly high speed is likely to be more readily apparent on the true motion display from the change in direction of the echo trail. True motion is generally more suitable for use with the lower range scales in congested waters rather than in the open sea. On vessels fitted with two radars in close proximity it may be advantageous to use the relative motion display on one radar and the true motion display on the other.

The following comment with regard to choice of display was made by the Netherlands Court at the inquiry into the collision between the vessels *Atys* and *Siena* (1963):

This collision teaches the following lesson with respect to the use of shipborne radar. The master declared he would have preferred to use his radar with true bearing presentation and on the 3 mile range scale. However he complied with the pilot's request and switched to the ship's head up display and the 1 mile range scale. In this particular case the master was right. Under the prevailing conditions it was misleading and dangerous to use the radar's 1 mile range scale in the congested approaches to the New Waterway. With the radar switched to the true motion presentation, it would have been much easier and faster to accurately determine the behaviour of the Siena. Determination of ships' movements by the observation of echoes on the radar display is much easier when a north up stabilised or true display is used than when a ship's head up or unstabilised display is used.

More generally speaking it can be said that under similar circumstances masters should not leave the decision of how the radar should be used and what presentation or range scale should be chosen, to the pilot. Especially since the shipborne radar can for these masters be a valuable aid for the proper navigation and conning of the vessel and can help them in judging the value of the pilot's advice.

The UK Government has issued a Marine Guidance Note (MGN 379) relating to the proper use of radar, including ARPA. This notice draws attention to the need for shipmasters and others using radar to gain and maintain experience in radar observation and appreciation by practice at sea in clear weather so that they can deal rapidly and competently with the problems which will confront them in restricted visibility.

In an action brought before the US District Court for the Eastern District of Pennsylvania in 1988 the owners of the *Seapride II* sought to limit their liability after the vessel struck a tower in the Delaware River. It was held that the owners were not entitled to limit their liability because the ship's master was not properly trained in the use of ARPA. The following comment was made with respect to the need for ARPA training:

The evidence clearly showed that both Captain Siderakis and Pilot Teal were not properly trained on ARPA. Petitioners may not have significant authority or control over Pilot Teal due to his compulsory and temporary tenure on the ship. They do, however, have the ability and responsibility to assure that its ship's master is sufficiently trained on the ship's equipment, particularly those devices required by law. It undermines the law that requires the equipment if shipowners fail to train their masters in its operation and use. Ironically the ship's management had a policy that required masters to teach ARPA skills to officers when it did not properly train the masters in the first place.

Plotting or equivalent systematic observation

Even continuous observation by a competent person is unlikely to be accepted as proper use of radar to obtain early warning of risk of collision if the bearings and distances of approaching vessels are not taken at regular intervals and carefully evaluated by plotting or by some equivalent method.

There are differences of opinion as to which form of manual plotting, either true or relative, should be used, but it is generally agreed that both methods have their advantages. The principles of relative plotting should be understood by all observers, as this is the method which enables the closest position of approach to be determined. The true plot is simpler to understand, and is considered by many to be superior when there are several targets on the screen. Alterations of course, or speed, by the observed vessel, carried out simultaneously with, or shortly after, an alteration by own vessel, are likely to be more readily detected by a true plot than by a relative plot.

The term 'equivalent systematic observation' would apply to the use of one or more of the various types of radar aids which are available, ranging from simple plotting devices to automatic radar plotting aids – ARPA (see page 21). It may even be accepted as applying to the recording of ranges and bearings at regular and frequent intervals by an observer who has no such aids at his disposal and who may find it impracticable to plot because of the particular circumstances.

In regions of high traffic density it may be impracticable to make, and to evaluate, a comprehensive manual plot, but, provided the radar is being carefully and continuously observed, it should be possible to discard some targets which are obviously going to pass well clear and concentrate on those with which a close quarters situation seems likely to develop.

Assumptions on scanty information

The determination of risk of collision, in both clear and restricted visibility, must be based on several successive observations taken as accurately as possible. The table on page 169 gives the change of bearing for each change of range of 1 mile as a vessel approaches or recedes, assuming that neither vessel alters course or speed. It will be seen that even if the closest approach distance is over 1 mile the rate of change of bearing will be relatively small at the longer ranges indicated. Small errors in ranges and bearings taken in the early stages of an encounter, or inaccurate plotting, are therefore likely to have an appreciable effect on the assessment of risk of collision.

Consider, for example, a target approximately a point on the starboard bow, the actual bearing of which remains constant. If bearings are taken when at distances of 12 miles and 10 miles away, and an error of $-1°$ is made in the first bearing followed by an error of $+1°$ in the second, the target would appear from the plot to be on a parallel and opposite course, and might be expected to pass clear to starboard with a nearest approach of over 2 miles. On the other hand, if an error of $+1°$ is made in the first bearing and $-1°$ in the second, it would appear that the target is crossing and should pass clear to port at a distance of over 2 miles.

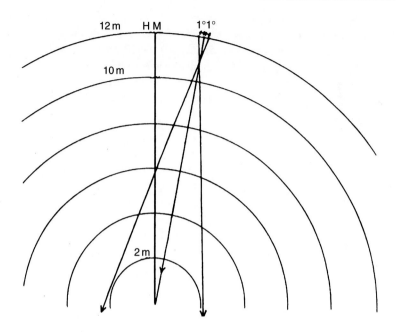

The example serves to illustrate the danger of attempting to assess whether risk of collision exists from a small number of observations taken at long range. Errors in bearings of ±1° are not unlikely when using radar and small errors in range measurement may also occur. Further inaccuracy may result from the reading and plotting of ranges and bearings. Several observations should be taken at short and regular intervals to reduce the effects of these random errors when there is a possibility of a close quarters situation developing.

Evje–Dona Evgenia

Basing what I am going to say upon those facts, I am unable to accept the evidence of the Evje *that initially the vessels were on opposite and parallel courses. Accurate observations and plottings by the* Evje *should, in my view, have revealed that the* Dona Evgenia *was in fact on a course which would lead her across the line of advance of the* Evje *from her port to her starboard bow. This initial error on the part of the* Evje *influenced her actions in relation to the* Dona Evgenia *at a comparatively early stage, because, wrongly supposing that the* Dona Evgenia *would pass her fairly closely to port, she starboarded in the manner I have described, expecting thereby to pass the* Dona Evgenia *with ample clearance to port. A moment's reflection upon her navigation at this time is sufficient to satisfy me that, by making such alterations to starboard in the circumstances I have just described, and in reducing her speed in the manner and at the times I have mentioned, she was in fact putting and maintaining herself on a series of courses (which for this purpose I may describe as an irregular arc) which brought her perilously near the line of advance of the* Dona Evgenia. *It is little wonder therefore, that, after a substantial alteration of course to starboard the master of the* Evje *found that the bearing of the* Dona Evgenia *had not changed as expected between the time he first began to starboard and the time he completed his turn.*

These facts, and the inferences which I have drawn from them, indicate a lamentable lack of appreciation of the situation which was developing all the time and which would have been obvious if accurate and simple plotting had been resorted to. (Mr Justice Hewson, 1960)

When vessels are in sight of one another visual compass bearings should normally give greater accuracy than radar bearings, but if the vessel is rolling or pitching heavily errors may be present, especially with the magnetic compass. Bearings taken relative to the ship's structure can be very misleading in determining whether risk of collision exists. Assumptions made on the basis of scanty information have been a contributory cause of many collisions in both clear and restricted visibility.

In the case of *Toni-Cardo* (1972) it was established that the *Cardo* altered course to starboard when the *Toni* had approached to within 5 miles, fine on the port bow, because the radar bearing seemed to be opening and the two vessels were expected to pass closely, port to port. The visibility was excellent and the navigation lights of the *Toni* had been seen at a considerable distance but no visual compass bearings were taken. The Judge concluded that the master of the *Toni* was keeping a poor look-out and that the *Toni* had altered course to port. It seems probable that the two vessels would have passed starboard to starboard if they had both kept their course and speed and that the *Toni* turned to port to increase the passing distance. Both vessels were found equally to blame.

No appreciable change of compass bearing

Risk of collision shall be deemed to exist if the compass bearing of an approaching vessel is not appreciably changing, not the relative bearing. The relative bearing will be affected by changes of heading. Sighting an approaching vessel against components of the ship's structure may give a rough indication of whether there is risk of collision and may provide sufficient basis for deciding whether to make a bold alteration to pass astern of a vessel being overtaken or crossing from the starboard side. Such bearings, however, must always be related to the ship's heading and may be affected by slight changes in the observer's position unless careful transits are taken.

Risk associated with changing bearing

When two vessels pass close to one another without any changes of course and speed the bearing which subtends from the other will remain almost constant at long range and change rapidly at short range (see table on page 173). An appreciable change of bearing at short range may therefore be associated with a dangerously close passing distance. The bearing will change by more than 5° as the range closes from 2 miles to 1 mile if the nearest approach is only 0.1 miles or 200 metres. Such a passing distance will bring danger of collision, especially if the vessels are in an overtaking situation in relatively shallow water (see pages 71–73).

An appreciable change of bearing at greater ranges does not necessarily mean that there is no risk of collision. The other vessel may be making a series of small alterations which have not been observed. This would apply especially in restricted visibility when the other vessel is being observed by radar only, but it may also apply when vessels are in sight of one another. The diagram illustrates a situation of this kind. Both ranges and bearings should be taken into account when determining whether there is risk of collision.

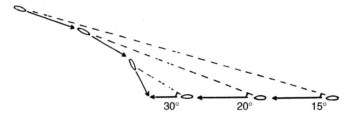

In the case of *Crystal Jewel–British Aviator,* 1964, the *British Aviator* observed the echo of the *Crystal Jewel* to broaden slowly from 9° on the starboard bow at a distance of 9 miles until the last reported bearing which was, reputedly, 45° on the starboard bow at ³⁄₄ of a mile. It was assumed that the other vessel was passing clear, starboard to starboard. The *Crystal Jewel* had, in fact, made a number of small alterations to starboard, anticipating a port to port situation.

RULE 8

Action to avoid collision

(a) **Any action to avoid collision shall be taken in accordance with the rules of this Part and, if the circumstances of the case admit, be positive, made in ample time and with due regard to the observance of good seamanship.**

(b) **Any alteration of course and/or speed to avoid collision shall, if the circumstances of the case admit, be large enough to be readily apparent to another vessel observing visually or by radar; a succession of small alterations of course and/or speed should be avoided.**

(c) **If there is sufficient sea room, alteration of course alone may be the most effective action to avoid a close-quarters situation provided that it is made in good time, is substantial and does not result in another close-quarters situation.**

(d) **Action taken to avoid collision with another vessel shall be such as to result in passing at a safe distance. The effectiveness of the action shall be carefully checked until the other vessel is finally past and clear.**

(e) **If necessary to avoid collision or allow more time to assess the situation, a vessel shall slacken her speed or take all way off by stopping or reversing her means of propulsion.**

(f) (i) **A vessel which, by any of these rules, is required not to impede the passage or safe passage of another vessel shall, when required by the** circumstances of the case, take early action to allow sufficient sea room for the safe passage of the other vessel.

(ii) A vessel required not to impede the passage or safe passage of another vessel is not relieved of this obligation if approaching the other vessel so as to involve risk of collision and shall, when taking action, have full regard to the action which may be required by the rules of this Part.

(iii) A vessel the passage of which is not to be impeded remains fully obliged to comply with the rules of this part when the two vessels are approaching one another so as to involve risk of collision.

COMMENT:

(a) Positive action in ample time

The circumstances must obviously be taken into account in considering what is meant by 'ample time'. In both clear and restricted visibility the situation should, if possible, be carefully assessed before action is taken. Assumptions should not be made on the basis of scanty information (see pages 32–33).

When vessels are in visual sight of one another the vessel which is directed to keep out of the way, must, so far as possible, take early avoiding action as required by Rule 16. If the give-way vessel takes action in good time the stand-on vessel will be required to maintain her course and speed and will not be justified in taking action in accordance with Rule 17(a)(ii).

An amendment to paragraph (a) was adopted by the 22nd Assembly of IMO, by which a direct link is established between Rule 8 on Action to Avoid Collision and the other Steering and Sailing Rules of Part B.

The reason for this amendment was that reports of collision cases indicated that at times in head-on, near head-on encounters or in fine crossing situations Rule 8 and in particular Rule 8 (d) was applied in isolation of the other Steering and Sailing Rules, resulting in conflicting actions and collisions.

In December 2002 IMO issued a Safety of Navigation Circular (SN Circular 226) on *DANGERS OF CONFLICTING ACTION IN COLLISION AVOIDANCE* explaining the above mentioned reason for the amendment of Rule 8(a).

'Conflicting actions may occur in head-on or near head-on encounters where one ship takes avoiding action by turning to port and the other ship by turning to starboard.

In investigations of collision cases the turn to port was explained to achieve a safe passing distance in accordance with Rule 8(d). The ship which took the avoiding action by turning to port ignored the possibility of initiating a conflicting action. An avoiding action to starboard by the approaching ship, in accordance with the other Steering and Sailing Rules in Sections II and III of Part B, was not anticipated.

The collision which occurred off the coast of South Africa in 1977 between the vessels Venoil *and* Venpet *is an example of conflicting action in a head-on situation. The vessels were approaching each other on reciprocal courses in restricted visibility.* Venoil *made a series of small alterations of course to starboard to increase the port-to-port passing distance.* Venpet *made small alterations of course to port to increase the starboard to starboard passing distance.'*

Maloja II–John M

The structure of the Collision Regulations is designed to ensure that, whenever possible, ships will not reach a close-quarters situation in which there is risk of collision and in which decisions have to be taken without time for proper thought. Manœuvres taken to avoid a close-quarters situation should be taken at a time when the responsible officer does not have to make a quick decision or a decision based on inadequate information. Those manœuvres should be such as to be readily apparent to the other ship. The errors of navigation which I regard as the most serious are those errors which are made by an officer who has time to think. At such time there is no excuse for failure to comply with the Collision Regulations. (Mr Justice Sheen, 1993)

Rule 8(a) does not give a vessel which is initially required to keep her course and speed the right to take action at an early stage. Rule 17(a)(ii) only permits such a vessel to manœuvre when it becomes apparent that the give-way vessel is not taking appropriate action. The stand-on vessel would probably not be justified in taking action to avoid collision before giving the whistle signal prescribed in Rule 34(d) (see page 81).

(b) Large enough to be readily apparent

Paragraph (b), combines, in mandatory form, Recommendations 7 and 5(a) of the Annex to the 1960 Rules but the application is extended to vessels in sight of one another which may not be using radar. The phrase 'if the circumstances of the case admit' is incorporated in the text to provide for situations in which large alterations cannot be made due to lack of sea room or other causes.

The need for substantial action has often been stressed in the Courts, for vessels in sight of one another as well as for vessels in restricted visibility.

Billings Victory–Warren Chase

I do not think it really needs repetition, because it has been said over and over again in this Court that the duty of the give-way vessel is to take timely action to keep clear. Moreover, it is her duty to act, if I may use the expression, handsomely, so as to leave the stand-on vessel in no possible doubt as to what the give-way vessel is doing. If her method of giving way is to alter course, she ought to make a substantial alteration, and ought to be particularly careful to signify that alteration by the appropriate helm signal. It may be said in this case that she could not safely make such an alteration of course because she was embarrassed by the presence of the pilot cutter. In that event, she would choose another method of keeping clear, by making a drastic reduction of her speed. Again, a sufficient alteration in her speed would leave the stand-on vessel in no doubt as to what she was doing. (Mr Justice Willmer, 1949)

In restricted visibility alterations of course and speed should be substantial so that they may be readily apparent to another vessel observing by radar. An alteration of course should be at least 30° for this purpose, and should preferably be of the order of 60° to 90°.

Alterations of speed take longer to put into effect than alterations of course so they are less likely to be readily observed. If a reduction of speed is to be made the way should be taken off as rapidly as possible by stopping the engines. Slow ahead or dead slow ahead can be ordered subsequently.

When vessels are in sight of one another it will probably be sufficient to make alterations of course which will be readily apparent to a person observing visually from the other ship. The Rule refers to 'another vessel observing visually or by radar'. Alterations of course of less than 10° are unlikely to be accepted as satisfying this requirement. A giving-way ship which alters course to pass astern of the other vessel should preferably turn sufficiently to bring the other vessel on to the opposite bow, so that at night a different sidelight would be visible, then gradually turn back maintaining the same relative bearing, until the original course is resumed.

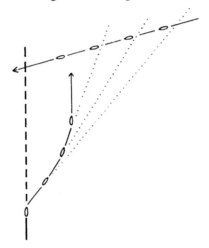

A common factor of many collisions involving vessels using radar has been the number of small alterations made by one, or both, of the vessels concerned. Small alterations are unlikely to be detected by the other vessel and may increase the danger of collision. In the case of the *British Aviator–Crystal Jewel* 1964 (referred to on page 35) the *British Aviator* failed to appreciate the series of small course alterations made by the *Crystal Jewel,* and considered that as the bearing was changing there was no risk of collision.

(c) Alteration of course alone

The distance at which a close quarters situation first applies will depend upon a number of factors, including the visibility.

In the open sea distances of the order of 2 to 3 miles are usually considered as the outer limits in restricted visibility but smaller distances, probably of the order of 1 mile, would probably be accepted for vessels in sight of one another (see also page 97).

An alteration of course will be more effective than a change of speed in order to avoid a vessel which is ahead or fine on the bow and this will also apply if action has to be taken to avoid an overtaking vessel approaching from astern or fine on the quarter which fails to keep out of the way. A change of speed is more effective than an alteration of course in order to avoid a vessel approaching from abeam or near the beam, but an alteration of course can be made to achieve the same result as a reduction of speed provided it is substantial. The diagram shows the effectiveness of turning over 60° to port to avoid a vessel approaching on a constant bearing from the starboard beam. It is usually safer to turn away from a vessel approaching from that direction (see also pages 174–176).

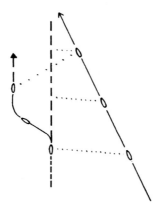

(d) Passing at a safe distance

The requirement that action taken to avoid collision shall be such as to result in passing at a safe distance is introduced for the first time in the 1972 Regulations. When vessels are in sight of one another, and one of two vessels is required to keep out of the way, the obligation to take action which will result in passing at a safe distance will obviously apply almost exclusively to the give-way vessel.

The stand-on vessel must initially keep her course and speed, and is only permitted to take action by Rule 17(a)(ii) when it becomes apparent to her that the vessel required to keep out of the way is not taking appropriate action. The first moment for such permitted action may not be at a sufficiently early stage to ensure that her manœuvre alone will achieve a really safe passing distance. It is therefore unlikely that the requirement of Rule 8(d) will be considered to apply to action permitted by Rule 17(a)(ii). The requirement could also hardly apply to action taken by the stand-on vessel in accordance with Rule 17(b) when the vessels are so close that collision cannot be avoided by the action of the give-way vessel alone. This is implied by the wording of the Rule which says that the stand-on vessel shall take such action as will best *aid* to prevent collision. The circumstances of the case must be taken into account.

In restricted visibility every vessel which detects the presence of another vessel by radar is required to take avoiding action if a close quarters situation is developing and/or risk of collision exists (Rule 19(d)), but the circumstances may not permit action to be taken which will result in passing at a safe distance. If, for instance in the open sea, a ship is detected ahead or fine on the bow and careful plotting or equivalent method of assessment indicates that the other vessel is proceeding at a relatively high speed, and that if no action is taken the two vessels will pass starboard to starboard at too close a distance, of the order of 1 mile, it may be dangerous to alter course either to starboard or to port. A reduction of speed may be the safest form of action in such circumstances but this may not appreciably affect the passing distance.

Effectiveness of action to be checked

As risk of collision is deemed to exist if the compass bearing of an approaching vessel does not appreciably change, a definite and continuing change of compass bearing would be one indication of the initial effectiveness of the avoiding action. However, an appreciable change of bearing may not be sufficient to establish that the vessels will eventually pass clear of one another. Subsequent action by the other vessel could result in renewed risk of collision. The situation could become even more dangerous than before if both vessels turn towards each other when crossing at a broad angle as the speed of approach may be increased.

The need to check the effectiveness of action taken to avoid collision applies especially in restricted visibility as subsequent action by the other vessel is less likely to be apparent on the radar screen. Plotting, or equivalent systematic observation, should therefore be continued until the other vessel is well clear.

When action is taken which could conflict with the action which is likely to be taken by the other vessel particular care should be taken. If, for instance, speed is reduced to avoid a vessel crossing from the port bow, a careful watch should be kept to see if the other vessel turns to starboard as this would probably necessitate a return to the original speed.

(e) Reductions of speed

Rule 8(e) must be interpreted in context with Rules 6, 19(b) and 19(e). Every vessel is required to proceed at a safe speed at all times. Although increases of speed, as a means of avoiding collision, are not prohibited, the emphasis in the Rules is placed on reductions of speed. As vessels infrequently proceed at a lower speed than would be considered safe for the prevailing circumstances an increase in speed large enough to satisfy the requirements of Rule 8(b) would usually be in contravention of Rule 6.

When a vessel is obliged to take action to avoid collision with another vessel which is crossing, or which she is overtaking, she may be prevented from making course alterations due to lack of sea room or to the presence of other vessels; in such circumstances it will be necessary to slacken speed or take all way off. In restricted visibility when a close quarters situation cannot be avoided with a vessel forward of the beam, or a fog signal is heard forward of the beam, it will usually be necessary to reduce speed or stop the ship (see pages 99–101).

The speed must also be reduced if it is necessary to allow more time to assess the situation. Rule 5 requires that a full appraisal of the situation and of the risk of collision should be made. When a vessel is sighted at short range and it is not possible to determine how she is heading due to poor visibility or weak lights the best action for a ship with a good stopping power may be to make a drastic reduction of speed.

In the *Buccleuch–Kyanite* 1905, the *Kyanite* altered course away from the danger when the loom of a sailing vessel was seen fine on the bow. Lord Low said:

But at that time... [the officer in charge of the Kyanite] *did not know that the* Buccleuch *was a crossing ship. He had no idea in what direction she was sailing. All that he knew was that a ship under sail was in dangerously close proximity. In such circumstances, I think that his duty was to stop and reverse. That was the one course which, I think, he could not have been wrong in following. What he did do was fatal, if, as it turned out, the* Buccleuch *was a crossing ship.*

Officers aboard modern power-driven vessels are usually reluctant to use the engines when it becomes necessary to keep out of the way of another vessel in the open sea. If the engines are not controlled from the bridge there is likely to be an appreciable delay before telegraph orders are put into effect unless the engines are on stand-by. Even if there is an immediate response a large vessel moving at high speed carries considerable momentum and cannot be expected to rapidly lose her way. If the engines are stopped on a tanker of over 200,000 tons deadweight it may take more than 20 minutes before the speed is halved and over an hour before the vessel comes to rest. These times can be considerably reduced by putting the engines astern as soon as possible but there is usually a delay of several minutes before the astern power becomes effective.

A drastic reduction of speed will be less readily apparent to another vessel than a substantial alteration of course, whether observed visually or by radar. When proceeding at full speed most vessels are capable of turning through at least 60° in the first 2 minutes if full helm is applied. Helm action will also cause the speed to be reduced.

Although helm action is usually preferable to engine action as a means of avoiding collision in the open sea, the officer of the watch should not hesitate to use the engines if the necessity arises (see page 152). The engines can normally be used to greater effect for collision avoidance when proceeding at reduced speed with the engines ready for immediate manœuvre, in restricted visibility or within port limits.

Taking all way off

If the engines are stopped on a vessel proceeding at high speed, or if ahead power is substantially reduced, there will be a fairly sharp drop in speed at first followed by a more gradual decrease, as hull resistance may be considered to be proportional to the square of the speed. If the engines are put astern shortly afterwards the initial high decrease of speed will be maintained and headway will fall off even more rapidly when the astern power becomes fully effective. The graph shows the fall off of speed against time as observed on three vessels when the engines were stopped while proceeding at full speed. The displacements were: vessel A 22,000 tons, vessel B 56,000 tons and vessel C 240,000 tons. The dotted lines indicate the effect of putting the engines astern at the earliest possible moment.

There will usually be a delay of at least one to two minutes before the engines can be reversed after moving at full head, depending upon the type of machinery and other factors. The maximum astern power is likely to be less than maximum ahead power.

For diesel engines the proportion will usually be over 80% but for geared steam
turbines it may only be about 40%.

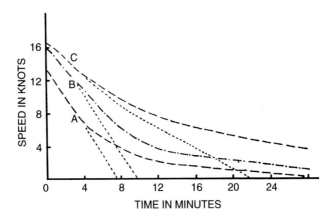

There have been several instances of vessels having their engines immobilised due
to attempting to reverse too rapidly after going at full ahead. In a paper presented
before the Institute of Marine Engineers in 1957 Mr J. E. Church described the
damage which could result to different types of machinery through a crash stop
manœuvre and suggested that a quicker and safer way to stop a vessel would be to
stop the engines instantly then, after a delay of three minutes or so, to give slow
astern, half astern and full astern, thus avoiding acute cavitation. More recent
evidence from ship trials and model tests seems to indicate that many vessels,
especially those fitted with diesel engines, could best be stopped by giving 'full
astern' as soon as possible, but it can generally be said that even if the engines can be
made to go astern within one minute of the order 'stop' the retarding effect would be
small and the risk of damage to the machinery would be great. The above remarks
apply to a vessel moving at high speed. The engines can be more readily reversed
when the speed is low.

The distance that a vessel will cover in a crash stop before being brought to rest
from full speed is likely to be between 5 and 15 ship lengths according to speed,
displacement, type of engine, etc. The time taken will vary considerably. A cargo
vessel of 3,000 tons displacement proceeding at 16 knots has been stopped in less than
3 minutes, but it may take more than 25 minutes to stop a loaded tanker of over
200,000 tons displacement moving at the same initial speed.

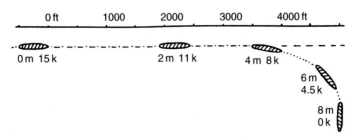

The diagram shows the path traced out by a vessel of about 23,000 tons displacement when making a crash stop after going at her full speed of approximately 15 knots. The rudder will begin to lose its effect as soon as the engines are stopped and it will become increasingly difficult to steer the vessel as astern power is developed. With a right-handed single screw ship the effect of the propeller when going astern on the engines is usually to slew the stern round to port so that by the time the vessel has been brought to rest in a crash stop she may have come off her course by 90° or more. If a strong wind is blowing this may have a greater effect on a vessel in light conditions than the transverse thrust of the propeller. Shallow water effect may also influence the vessel's heading during a crash stop.

The above comments apply mainly to the stopping of a vessel with a right-handed single screw conventional propeller. In Rule 8(e) it is stated that, if necessary, a vessel shall 'take all way off by stopping or reversing her means of propulsion'. The wording takes account of the fact that controllable pitch propellers are being fitted to an increasing number of ships. It will normally be possible to stop a vessel more rapidly with reduced head reach if a controllable pitch propeller is fitted. The most effective method of taking off the way when using a propeller of this type may be to gradually change the pitch as the speed is reduced so as to give maximum reverse thrust but there is some difference of opinion on this point.

Helm action taken in the initial stage of a crash stop, when still moving at high speed, will result in a considerable increase of resistance and reduce the stopping distance. A method which may be used in some circumstances is to put the helm hard over one way then hard over to the other side with the engines on dead slow ahead, then to put the engines full astern. This should reduce the period of applying astern power so that the vessel is less likely to be slewed in the final stage.

Sailing vessels

Rule 8(e) also applies to sailing vessels. Sailing vessels must, if necessary, slacken speed in order to avoid collision or allow more time to assess the situation. This can be achieved by luffing up into the wind or reducing sail. Moving the rudder hard over to one side then hard over to the other may also help to bring the speed down.

(f) Not to impede

Rule 8(f) was first adopted by IMO in 1987 to cover the Guidance for the uniform application of the words 'not to impede' which appear in Rules 9(b), (c) and (d), 10(i) and (j) and 18(d). The following item of Guidance was approved in 1982 by the Maritime Safety Committee;

'When a vessel is required not to impede the passage of another vessel, such a vessel shall so far as practicable navigate in such a way as to avoid the development of risk of collision. If, however, a situation has developed so as to involve risk of collision, the relevant Steering and Sailing Rules shall be complied with.'

The above Guidance is superseded by Rule 8(f) which establishes clearly that the requirements of 'not to impede' are complementary to other requirements of the Steering and Sailing Rules.

The requirement not to impede the passage or safe passage of another vessel does not apply only to vessels in sight of each other which are approaching in such a way that risk of collision is likely to develop. The requirements of Rule 8(f) together with Rules 9(b), (c) and (d), and 10(i) and (j) apply in both clear and restricted visibility. For instance, a sailing vessel or small power-driven vessel which becomes aware of the approach of a large power-driven vessel which can safely navigate only within a narrow channel should take early action to allow safe passage whether or not the other vessel is in sight.

When vessels are in sight of each other and are approaching in such a way that risk of collision seems likely to develop the Rules of Part B Section II become applicable. In such circumstances a vessel which is required not to impede the passage of another vessel is not relieved of that obligation if the other vessel will become the give-way vessel when risk of collision exists. For instance, when a power-driven vessel and a sailing vessel are approaching each other the power-driven vessel is required by Rule 18(a) to keep out of the way when risk of collision begins to apply, although she may be proceeding along a narrow channel or traffic lane, but this does not relieve the sailing vessel of the obligation to take early action to allow sufficient sea room.

If one of two power-driven vessels, crossing so as to involve risk of collision, is required not to impede the passage of the other vessel, she must, in compliance with Rule 8(f), take early action to allow sufficient sea room for the safe passage of the other vessel although the other vessel may be required by Rule 15 to keep out of the way (see page 76).

It will not always be possible, in the circumstances of the case, for the vessel required not to impede to take early action to allow sufficient sea room for the safe passage of the other vessel. For instance, the day signal of a vessel constrained by her draught may not be recognised at sufficient distance for early action to be taken and restricted visibility may make it difficult to take early action in accordance with the relevant paragraphs of Rules 9 and 10.

Rule 8(f)(ii) establishes clearly that a vessel required not to impede does not lose that obligation if approaching the other vessel so as to involve risk of collision. Although the other vessel may become the give-way vessel when risk of collision develops the vessel required not to impede is not relieved from the requirement to allow sufficient sea room for the safe passage of the other vessel because of the application of Rule 17(a)(i). Early action in compliance with Rule 8(f) is compatible with Rule 17(a)(ii), which permits action by the stand-on vessel (see pages 80–81).

A vessel taking action so as to avoid impeding the passage of another vessel must have full regard to the action which may be required by the Steering and Sailing Rules. This is a requirement of Rule 8(f)(ii) to take account of the possibility of both vessels taking conflicting action when there is risk of collision. However, as it is not possible to establish the precise distance apart at which risk of collision begins to apply, a vessel taking early action not to impede should also have full regard to the action which may be taken by the other vessel. Rules 14, 15 and 17(c) indicate the form of action to be taken.

Rule 8(f)(iii) relates to the obligations of a vessel the passage of which is not to be impeded. Such a vessel is not relieved of her obligation to comply with the Steering and Sailing Rules when there is risk of collision. When vessels are in sight of one another and risk of collision exists, a power-driven vessel may be required to keep out of the way of the vessel required not to impede in accordance with Rules 13, 14, 15 and 18(a). In restricted visibility such a vessel is not relieved of her obligation to take avoiding action in ample time when a close quarters situation is developing. When there is an obligation not to impede in restricted visibility Rule 19 applies fully, together with Rule 8(f).

RULE 9

Narrow channels

(a) **A vessel proceeding along the course of a narrow channel or fairway shall keep as near to the outer limit of the channel or fairway which lies on her starboard side as is safe and practicable.**

(b) **A vessel of less than 20 metres in length or a sailing vessel shall not impede the passage of a vessel which can safely navigate only within a narrow channel or fairway.**

(c) **A vessel engaged in fishing shall not impede the passage of any other vessel navigating within a narrow channel or fairway.**

(d) **A vessel shall not cross a narrow channel or fairway if such crossing impedes the passage of a vessel which can safely navigate only within such channel or fairway. The latter vessel may use the sound signal prescribed in Rule 34(d) if in doubt as to the intention of the crossing vessel.**

(e) (i) **In a narrow channel or fairway when overtaking can take place only if the vessel to be overtaken has to take action to permit safe passing, the vessel intending to overtake shall indicate her intention by sounding the appropriate signal prescribed in Rule 34(c)(i). The vessel to be overtaken shall, if in agreement, sound the appropriate signal prescribed in Rule 34(c)(ii) and take steps to permit safe passing. If in doubt she may sound the signals prescribed in Rule 34(d).**

(ii) **This Rule does not relieve the overtaking vessel of her obligation under Rule 13.**

(f) **A vessel nearing a bend or an area of a narrow channel or fairway where other vessels may be obscured by an intervening obstruction shall navigate with particular alertness and caution and shall sound the appropriate signal prescribed in Rule 34(e).**

(g) **Any vessel shall, if the circumstances of the case admit, avoid anchoring in a narrow channel.**

COMMENT:

Narrow channels

The term 'narrow channel' is not easily defined. In deciding whether a particular stretch of water is or is not a narrow channel the Courts take into account the evidence as to the way in which seamen usually navigate the locality and the advice given by the Elder Brethren. A narrow channel need not be of any particular length and does not necessarily terminate at the last of the buoys or objects marking the channel. The narrow channel rule has been held to apply to the passage between two piers and to 100 metres (yards) outwards beyond the objects marking a harbour entrance. It was held not to apply to a recommended route between two buoys where vessels could have gone outside them in safety.

Passages approximately 2 miles wide have sometimes been considered narrow channels. In considering the passage between Duncansby Head and the Skerries in the Pentland Firth (*Anna Salen–Thorshovdi*, 1954) Mr Justice Willmer said:

For myself, I certainly see difficulties in applying the 'narrow channel' rule to a passage which is nearly four miles wide. I should hardly have thought that 'narrow' was the word to use for this passage, for it is not a particularly narrow passage.

In the *Faith I–Independence* (US Court, 1992) the passage between buoys at the entrance to Delaware Bay, approximately 1.2 miles wide, was held not to be a narrow channel but it was held that good seamanship and prudent navigation require that every vessel keep to starboard if safe and practicable.

Rule 9 will apply to any narrow channel connected with the high seas which is navigable by seagoing vessels unless there is an inconsistent local rule. It does not apply to lanes of traffic separation schemes although such lanes may be relatively narrow. Vessels using traffic separation schemes must comply with Rule 10.

Fairway

The term 'fairway' is generally used to refer to an open navigable passage of water, or the channel dredged and maintained by the port authority. The fairway has been considered to be the deep water channel which may be marked by pecked lines on the chart for use by large vessels (The *Crackshot,* 1949) whereas the term 'narrow channel' has been held to refer to the whole width of navigable water between the lines of buoys (*Koningin Juliana,* 1973).

Proceeding along the course of a narrow channel

A vessel is only required to keep near to the outer limits on her starboard side when proceeding along the course of the channel. She would, of course, be permitted to cross the channel for such purposes as changing pilots or proceeding to a side channel or berth which lies on the other side provided that such crossing does not impede the passage of a vessel which can safely navigate only within the channel (Rule 9(d)).

Sailing vessels proceeding along the course of the channel are required to keep to the starboard side so far as practicable. If it is not possible for them to keep close to the outer limit because of the direction of the wind they must comply with Rule 9(b).

Keep near to the outer limit

The requirement to keep near to the outer limit will usually mean that, when the depth of water diminishes from the mid-channel outwards to the sides, vessels with shallow draught must keep further to starboard than vessels of deeper draught. However, vessels are not expected to put themselves in danger by passing too close to the shoals, or to make frequent alterations of course in order to keep near to the outer limit of every part of the channel. They are required to keep as near to the outer limit as is safe and practicable.

It will not be sufficient to move over to the starboard side when encountering vessels proceeding in the opposite direction. A vessel is expected to *keep* near to the outer limit on her starboard side.

Use of radar

Vessels proceeding along the course of a narrow channel or fairway should make full use of radar and other navigational equipment, when necessary, to get to their correct side and to ensure that they are keeping as near to the outer limit as is safe and practicable. This will apply especially when the visibility is restricted. Several vessels have been criticised in the Courts for failing to use radar for this purpose.

British Tenacity–Minster

The Minster *was generously fitted with electronic navigational aids, and yet she failed to enter this narrow channel upon her proper side and failed at all times thereafter to get to it. (Mr Justice Hewson, 1963)*

(b) Small craft and sailing vessels

All vessels of less than 20 metres in length, and sailing vessels of any size, must avoid impeding the passage of a vessel which can safely navigate only within a narrow channel or fairway. This Rule differs from the Rules in Part B, Section II which generally allocate prime responsibility to one of two vessels approaching so as to involve risk of collision. In this case the application of Rule 8(f) is relevant (see pages 43–45).

Rule 8(f)(i) requires early action from a vessel which must avoid impeding the passage of another vessel, when such action is required by the circumstances of the case.

Small craft and sailing vessels are therefore expected to take early action to keep well clear of vessels which can only navigate within the channel or fairway, without waiting to determine if risk of collision exists. This Rule does not relieve a power-driven vessel which is restricted to the channel from her obligation to keep out of the way of a small power-driven vessel being overtaken or crossing from her starboard side, or of any sailing vessel, if there is risk of collision.

Small vessels and sailing vessels are not required to avoid impeding the passage of all power-driven vessels of over 20 metres in length when in a narrow channel. Many power-driven vessels which exceed that length may be able to navigate outside the channel limits. However, it is not only the passage of a vessel showing the lights or shapes to indicate that she is constrained by her draught which must not be impeded. Small vessels should take the width and depth of the channel into account and, if in doubt, keep clear of vessels likely to be restricted.

(c) Vessels engaged in fishing

Rule 9(c) requires that the passage of other vessels navigating within the channel should not be impeded. It implies that fishing is permitted when the channel is not being used.

Rule 9(c) has a wider application than Rule 9(b). It is not only vessels which can only navigate within the channel which must not be impeded. Even sailing vessels and small power-driven vessels must be allowed to pass along or across the channel or fairway. Fishing vessels are therefore only permitted to fish in a channel or fairway when they are able to get an early indication of the approach of other vessels which will enable them to clear the passage in sufficient time (see also pages 44–45).

(d) Crossing a narrow channel

Vessels must not cross a narrow channel or fairway if by doing so they impede the passage of any other vessel which can safely navigate only within the channel. However, this does not mean that the Crossing Rule does not apply in narrow channels. A power-driven vessel which can safely navigate outside the channel must keep out of the way of a power-driven vessel which is crossing the channel and is approaching from her starboard side so as to involve risk of collision. In a crossing situation, vessels should if necessary reduce speed in compliance with Rule 8(e) (see also page 40).

The main purpose of Rule 9(d) is to reduce the number of dangerous crossings in narrow channels or fairways, often caused by relatively small vessels which could usually avoid the danger by waiting until the passage is clear or by a better anticipation of the prevailing traffic situation. The application of Rule 8(f) is also relevant in this case (see pages 43–45).

If a vessel restricted to the channel is in doubt of the intentions of a vessel crossing the sound signal comprising of at least five short blasts should be given in compliance with Rule 34(d).

Vessels entering a channel

Rule 9 does not deal specifically with the case of a vessel entering a channel with the intention of proceeding along it but Rule 2(a) would probably apply in such circumstances as the following extracts from judgments indicate:

Canberra Star-City of Lyons

The rule of good seamanship for a vessel entering a main channel is that she should do so with caution and not hamper traffic already navigating in it. Vessels already in it, as well as those about to enter it, should behave reasonably. It does not appear to me that the vessel in the channel has a complete right of way, and she must not hog the river regardless of the reasonable aspirations of other vessels. (Mr Justice Hewson, 1962)

Burton-Prince Leopold de Belgique

There is no Rule, I understand, which applies to this particular point, and having discussed the matter with the Elder Brethren, as far as I can understand, vessels must deal with each other on the footing of good seamanship, of course complying, as far as possible, with the necessity of keeping on their starboard hand of the channel. It results from that, that if one vessel comes to the point of intersection reasonably in advance of the other, she must keep on, and the other must wait till she has passed. If both approach the spot at about the same time, then they must act reasonably, and it would be very reasonable that the one which has the tide against her should wait while the other passed. (Sir Gorell Barnes, 1908)

In the above case the two vessels were approaching on slightly crossing courses in the entrance channel at Swansea.

(e) Overtaking in a narrow channel

This paragraph was introduced for the first time in the 1972 Rules. The need for such a provision became increasingly apparent with the tendency of large vessels to proceed along a fairway or channel at high water and to overtake other vessels because of the limited time available. In such cases there is often insufficient room for overtaking to take place unless the vessel to be overtaken takes appropriate action to permit safe passing.

The procedure to be adopted when overtaking can only take place by mutual agreement is described in Rule 9(e)(i). On hearing a signal from the overtaking vessel indicating which side she intends to pass the vessel about to be overtaken should indicate agreement if it is safe to overtake and take such action as may be necessary to permit safe passing. It would be good seamanship to move away, as far as is safe and practicable, from the side of the fairway in which the overtaking vessel intends to pass, to allow a greater passing distance, and furthermore to reduce speed in order to decrease the period of running closely parallel to each other.

A vessel about to be overtaken must take account of the signals of intent made by the vessel wishing to overtake. If it is not considered safe for the other vessel to pass the signal of at least five short rapid blasts could be made on the whistle. This signal indicates doubt about the intentions or actions of the other vessel and implies that the vessel ahead does not consider it safe for the vessel astern to attempt to pass. In such circumstances the whistle signals should, if possible, be supplemented by the use of VHF radio communication to clarify the situation. The radiotelephone may also be used to advantage when it is considered safe to pass, in order to reach a clear understanding of the procedure to be followed.

Although Rule 9 is in Section 1 of Part B which relates to conduct in all conditions of visibility, the signals prescribed in Rule 34(c) are only to be made by vessels in sight of one another. As Rules 9(e) and 34(c) are complementary it is implied that Rule 9(e) applies only to vessels in visual sight of each other.

(f) Bends in a channel

This Rule applies to all vessels, not just to power-driven vessels, and is extended to cover areas of a narrow channel or fairway where other vessels may be obscured by an intervening obstruction.

On approaching a bend, or section of the channel where other vessels may be obscured, a vessel must sound one prolonged blast as prescribed in Rule 34(e). If, a few minutes later, a signal is heard from another vessel which is approaching the bend, this must be answered by a further signal of one prolonged blast.

The bend must be rounded with alertness and caution regardless of whether an approaching vessel is heard. A power-driven vessel must not 'cut the corner' and get on to the wrong side of the fairway. When two power-driven vessels approaching from opposite directions hear each other's signals it may be a precaution demanded by good seamanship for the vessel stemming the tide to wait until the other has passed clear.

Trevethick–Talabot

In the River Thames there is a well recognised and positive Rule (No. 23) of the Thames Conservancy to the effect that, when vessels are approaching each other, at bends such as I have described, it is the duty of the one having the tide against her to ease her engines and to wait under the point until the other vessel has passed it. There is, so far as appears from the present case, no such positive Rule printed and circulated with regard to the navigation of this Belgian river; but the pilots agree that the practice of navigation is really the same as that prescribed in the Rules for the navigation of the Thames. The vessel having the tide against her is to wait until the vessel coming in the opposite direction has cleared her at the bend. It is quite certain that the Trevethick *disregarded that practice, and I have no hesitation in saying that she was to blame for so doing. Nay, more; if there had been no such practice in the River Scheldt, both I and the Trinity Brethren are of opinion that it was bad navigation for the vessel with the tide against her to proceed as she did under the circumstances. (Mr Justice Butt, 1890)*

(g) Anchoring in a narrow channel

The requirement to avoid anchoring in a narrow channel, if the circumstances of the case admit, is new to the 1972 Rules. A vessel anchored in a narrow channel is likely to impede the safe passage of other vessels. Thick fog may not be considered justification for anchoring in a channel or fairway as it is common practice for many vessels to proceed with the assistance of radar. A vessel which finds it necessary to anchor in a narrow channel should endeavour to do so in a position where she will not obstruct the flow of traffic.

RULE 10

Traffic separation schemes

(a) **This Rule applies to traffic separation schemes adopted by the Organization and does not relieve any vessel of her obligation under any other rule.**

(b) A vessel using a traffic separation scheme shall:
  (i)   proceed in the appropriate traffic lane in the general direction of traffic flow for that lane;
  (ii) so far as practicable keep clear of a traffic separation line or separation zone;
  (iii) normally join or leave a traffic lane at the termination of the lane, but when joining or leaving from either side shall do so at as small an angle to the general direction of traffic flow as practicable.
(c) A vessel shall so far as practicable avoid crossing traffic lanes, but if obliged to do so shall cross on a heading as nearly as practicable at right angles to the general direction of traffic flow.
(d) (i) A vessel shall not use an inshore traffic zone when she can safely use the appropriate traffic lane within the adjacent traffic separation scheme. However, vessels of less than 20 m in length, sailing vessels and vessels engaged in fishing may use the inshore traffic zone.
   (ii) Notwithstanding subparagraph d(i), a vessel may use an inshore traffic zone when en route to or from a port, offshore installation or structure, pilot station or any other place situated within the inshore traffic zone or to avoid immediate danger.
(e) A vessel other than a crossing vessel or a vessel joining or leaving a lane shall not normally enter a separation zone or cross a separation line except:
  (i) in cases of emergency to avoid immediate danger;
  (ii) to engage in fishing within a separation zone.
(f) A vessel navigating in areas near the terminations of traffic separation schemes shall do so with particular caution.
(g) A vessel shall so far as practicable avoid anchoring in a traffic separation scheme or in areas near its terminations.
(h) A vessel not using a traffic separation scheme shall avoid it by as wide a margin as is practicable.
(i) A vessel engaged in fishing shall not impede the passage of any vessel following a traffic lane.
(j) A vessel of less than 20 metres in length or a sailing vessel shall not impede the safe passage of a power-driven vessel following a traffic lane.
(k) A vessel restricted in her ability to manœuvre when engaged in an operation for the maintenance of safety of navigation in a traffic separation scheme is exempted from complying with this Rule to the extent necessary to carry out the operation.
(l) A vessel restricted in her ability to manœuvre when engaged in an operation for the laying, servicing or picking up of a submarine cable, within a traffic separation scheme, is exempted from complying with this Rule to the extent necessary to carry out the operation.

Ships' routeing

The separation of opposing streams of traffic by means of traffic separation schemes is one of several routeing measures adopted by IMO. The IMO's responsibility for ships' routing is founded in the International Convention for Safety Of Life At Sea (SOLAS), 1974, as amended, Chapter V/Regulation 10, which recognizes the IMO as the only international body for establishing on an international level ships' routeing systems. The purpose of ships' routeing is to improve the safety of navigation in converging areas and in areas where the density of traffic is great or where the freedom of movement of shipping is inhibited by restricted sea-room, the existence of obstructions to navigation, limited depths or unfavourable meteorological conditions.

The criteria and principles applicable to all routeing measures are set out in the General Provisions on Ships' Routeing, which form part of IMO Resolution A 572(14) as amended. Routeing schemes and the General Provisions on Ships' Routeing are kept under continuous review by IMO and amendments are made when required.

The use of routeing systems

The following principles on the use of routeing systems are laid down in the IMO General Provisions on Ships' Routeing:

1. Unless stated otherwise, routeing systems are recommended for use by all ships and may be made mandatory for all ships, certain categories of ships or ships carrying certain cargoes.
2. Routeing systems are intended for use by day and by night in all weathers, in ice-free waters or under light ice conditions where no extraordinary manœuvres or assistance by ice-breaker(s) are required.
3. Bearing in mind the need for adequate under-keel clearance, a decision to use a routeing system must take into account the charted depth, the possibility of changes in the sea-bed since the time of the last survey, and the effects of meteorological and tidal conditions on water depths.
4. A ship navigating in or near a traffic separation scheme adopted by IMO shall in particular comply with Rule 10 of the 1972 International Collision Regulations to minimise the development of risk of collision with another ship. The other rules of the 1972 Collision Regulations apply in all respects, and particularly the rules of Part B, Sections II and III, if risk of collision with another ship is deemed to exist.
5. At junction points where traffic from various directions meets, a true separation of traffic is not really possible, as ships may need to cross routes or change to another route. Ships should therefore navigate with great caution in such areas and be aware that the mere fact that a ship is proceeding along a through-going route gives that ship no special privilege or right of way.
6. A deep-water route is primarily intended for use by ships which, because of their draught in relation to the available depth of water in the area concerned, require the use of such a route. Through traffic to which the above consideration does not apply should, as far as practicable, avoid using deep-water routes.
7. Precautionary areas should be avoided, if practicable, by passing ships not making use of the associated traffic separation schemes or deep-water routes, or entering or leaving adjacent ports.

8. In two-way routes, including two-way deep-water routes, ships should as far as practicable keep to the starboard side.

9. Arrows printed on charts in connection with routeing systems merely indicate the general direction of established or recommended traffic flow; ships need not set their courses strictly along the arrows.

10. The signal *YG* meaning *You appear not to be complying with the traffic separation scheme* is provided in the International Code of Signals for appropriate use.

Traffic separation schemes

Paragraph (a) of Rule 10 makes it clear that the Rule only applies to traffic separation schemes adopted by IMO (see below).

The words 'and does not relieve any vessel of her obligation under any other Rule' were added to Rule 10(a) by the 1987 amendment. This change was made to make it quite clear that all other Rules of the Collision Regulations continue to apply to a vessel using a traffic separation scheme. For instance, a power-driven vessel following a traffic lane is not relieved of her obligation to keep out of the way of a vessel seen on her own starboard side to be crossing so as to involve risk of collision.

As Rules 1(d) and 10(a) refer to traffic separation schemes adopted by the Organization it is implied that the terminology used in Rule 10 is the same as that included in the IMO General Provisions on Ships' Routeing. The terms 'traffic separation scheme', 'separation zone or line', 'traffic lane' and 'inshore traffic zone', which are used in Rule 10, are defined in the General Provisions on Ships' Routeing as follows:

(a) *Traffic Separation Scheme* A routeing measure aimed at the separation of opposing streams of traffic by appropriate means and by the establishment of traffic lanes.

(b) *Traffic Lane* An area within defined limits in which one-way traffic is established. Natural obstacles, including those forming separation zones, may constitute a boundary.

(c) *Separation Zone or Line* A zone or line separating the traffic lanes in which ships are proceeding in opposite or nearly opposite directions; or separating a traffic lane from the adjacent sea area; or separating traffic lanes designated for particular classes of ship proceeding in the same direction.

(d) *Inshore Traffic Zone* A routeing measure comprising a designated area between the landward boundary of a traffic separation scheme and the adjacent coast, to be used in accordance with the provisions of Rule 10(d), as amended, of the International Regulations for Preventing Collisions at Sea, 1972 (Collision Regulations).

Before its adoption by IMO, the description of each routeing system, routeing measure or route must be approved by the Maritime Safety Committee of IMO. A Traffic Separation Scheme may be described as a singular routeing measure, but also as part of a Routeing System. "Routeing System", as defined in the General Provisions on Ships' Routeing, is: "Any system of one or more routes or routeing measures aimed at reducing the risk of casualty, it includes traffic separation schemes, two-way routes, recommended tracks, areas to be avoided, no anchorage area, inshore traffic zones, roundabouts, precautionary areas and deep-water routes". In sea areas where routeing systems adopted by IMO are established Rule 10 exclusively applies to the part or parts described and adopted by IMO as traffic separation scheme(s). Rule 10 does not apply to other routes or routing measures, being part or parts of the same routeing system and connected to an adjoining traffic separation scheme.

Details of traffic separation schemes adopted by IMO are depicted on nautical charts, using the symbols which are described in the IMO publication *Ships' Routeing*. A government may however in urgent cases adjust an adopted scheme and implement this change before approval by IMO. It is important to keep charts up to date with respect to any new traffic separation schemes, or amendments to existing schemes, from information given in Notices to Mariners and other publications (see page 2).

Some governments have adopted, within their territorial waters, traffic separation schemes with principles and nomenclature that differ from those officially adopted by IMO. Mariners should consult nautical publications such as Sailing Directions and other relevant documents to see whether there are any important differences in the principles and nomenclature of a locally adopted scheme with which they should become familiar.

A government may also recommend the use of traffic separation schemes in international waters, without having submitted such schemes to IMO for adoption. Rule 10 will not apply to such schemes but it may be considered good seamanship to comply with the recommendations relating to their use in accordance with Rule 2(a). Off the coast of Japan several traffic separation schemes are recommended for use by the Japanese Captains' Association since 1970, but are not adopted by IMO. In 1973 a collision occurred in a traffic lane of one of these schemes between the *American Aquarius* and the *Atlantic Hope*. It was held in the United States Court of Appeals that the traffic separation scheme had not attained the status of a custom and that the action of the *American Aquarius* in proceeding in the wrong direction in the traffic lane could not be fairly characterised as a failure to conform with good seamanship.

Vessels using a traffic separation scheme

Paragraph (b) applies to vessels using a traffic separation scheme and paragraph (h) applies to vessels not using a traffic separation scheme. A vessel is using a scheme, in the context of Rule 10, when she is navigating within the outer limits of the scheme and is not crossing the lanes nor engaged in fishing within a separation zone. A vessel using an inshore traffic zone is not using the scheme.

Any vessel using a traffic separation scheme, including a sailing vessel, would normally be required to proceed in the appropriate traffic lane in the general direction of traffic flow. However, it is clearly necessary to permit essential activities, such as hydrographic surveying, to take place within the area covered by a traffic separation scheme. Among the amendments to the Rules, adopted by the IMO General Assembly in 1981, are two additional paragraphs to Rule 10 which provide for this need. A vessel which is engaged in an operation for the maintenance of safety of navigation, or in the laying, servicing or picking up of a submarine cable, within a traffic separation scheme is exempted, by paragraphs (k) and (l), from complying with Rule 10 to the extent necessary to carry out such work, if she is restricted in her ability to manœuvre. Such a vessel is, therefore, not prohibited from proceeding against the general direction of flow within a traffic lane if this becomes necessary to carry out the operation.

There is no exemption from complying with Rule 10 for a vessel engaged in fishing. Although fishing is not prohibited within a traffic lane a vessel engaged in fishing is not permitted to proceed in the opposite direction to the general direction of traffic flow (see pages 59–60).

The general direction of traffic flow within a traffic lane is indicated by arrows on the charts which are usually staggered so as to avoid the suggestion of a preferred track (see page 52). This information may also be given in the IMO publication *Ships' Routeing*.

A vessel which proceeds in the wrong direction in a traffic lane considerably increases the risk of collision and is likely to be found seriously at fault if a collision should occur. Vessels which violate Rule 10 may also be reported to their respective governments and prosecutions may ensue.

A vessel proceeding along a traffic lane in the direction of flow is required to keep clear of a zone or line separating traffic proceeding in opposite directions. As the boundaries of traffic lanes are not usually marked by buoys and it is not always possible to determine position in a lane with a high degree of accuracy there is a danger that a vessel which sets a course near the edge of a lane will stray into the separation zone or the traffic lane designated for traffic proceeding in the opposite direction. This requirement is intended to give greater effect to the separation of opposing streams of traffic.

It is also important that a vessel proceeding along a traffic lane should keep clear of the outer limit which lies on her starboard side, particularly if this line separates the lane from an inshore zone which is likely to contain traffic moving in the opposite direction. on the edge of the lane two power-driven vessels meeting on reciprocal courses would each be required to alter course to starboard by Rule 14. Such actions may cause both vessels to be involved in further meeting situations making it difficult for them to return to their correct lane or zone.

Paragraph (b)(iii) was amended in 1981 so that the requirement will apply to vessels joining or leaving a lane from either side. The amendment makes it clear that a vessel which crosses one traffic lane before joining, or after leaving, the other one is expected to join, or leave, at a small angle to the direction of traffic flow.

Crossing traffic lanes

A vessel must avoid crossing lanes, so far as practicable. In some areas covered by a traffic separation scheme, such as the Dover Strait, it is not possible for ferries and other vessels to avoid crossing the lanes without making a considerable detour so there is an appreciable amount of crossing traffic. There can be little justification for crossing the relatively short traffic lanes which have been established off capes and headlands in many areas.

Crossing a traffic lane may disturb the traffic flow pattern and increase the risk of collision. When risk of collision exists between vessels in a traffic lane the relevant Rule from Part B will usually apply, although small vessels and sailing vessels have a general obligation to avoid impeding the safe passage of power-driven vessels following the lane. A power-driven vessel, proceeding along a traffic lane, which sees another power-driven vessel crossing from her own starboard side will usually be required to keep clear by Rule 15. It may be difficult for such a vessel to take substantial avoiding action without making it necessary for other vessels to manœuvre.

The requirement to cross as nearly as practicable at right angles to the general direction of traffic flow applies at all times, whether there are vessels proceeding along the lane or not.

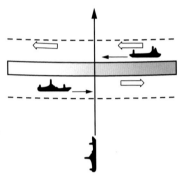

The words 'on a heading' were added to Rule 10(c) by the 1987 amendments to make it clear that it is the heading of the vessel and not the course made good which should be as nearly as practicable at right angles to the direction of traffic flow. For slow vessels experiencing a strong cross current or tidal stream there can be an appreciable difference between the course steered and the course made good. The shortest time to cross a traffic lane is achieved by crossing on a heading at right angles to the direction of traffic flow.

A vessel will only be justified in crossing a lane at an angle which differs appreciably from 90° if there are special circumstances such as the need to keep clear of another vessel or severe weather conditions. A sailing vessel may be unable to cross at right angles because of the direction of the wind but an auxiliary engine, if fitted, should be used in order to cross as nearly as practicable at right angles (see Marine Guidance Note MGN 364 issued by the Government of the United Kingdom).

In August 1986 the sail training vessel *De Eendracht*, fitted with an auxiliary engine, was proceeding under sail only on a passage from Heligoland to Terschelling. Off Terschelling she crossed the southern traffic lane of the traffic separation scheme on a course of 192°. As the general direction of traffic flow in the southern lane is 072° a vessel proceeding towards Terschelling is required to cross on a heading as nearly as practicable to 162°. The officer of the watch of the *De Eendracht* decided to cross on a heading of 192° to avoid the risk of uncontrolled gybing because the wind direction was approximately 340°.

The officer of the watch of the *De Eendracht* was subsequently prosecuted in a Court in Amsterdam and was found guilty of contravening Rule 10(c) because he had failed to use the auxiliary engine to achieve a right-angled crossing.

Where traffic is under surveillance by shore radar equipment the controlling authorities should make due allowance for the effect of tide, current or wind in assessing whether a vessel is crossing a traffic lane as nearly as practicable at right angles to the direction of traffic flow.

In March 1976 the IMO sub-committee on Safety of Navigation considered the problem of French fishing vessels crossing the traffic lanes in the Dover Strait. The sub-committee came to the conclusion that it is within the ordinary practice of prudent seamanship and in keeping with the provisions of Rule 10 of the 1972 Regulations that a vessel using a traffic lane may make a transfer within a lane from one side to the other, provided such a transfer is accomplished at as small an angle to the general direction of traffic flow as is practicable. This point has now been incorporated in the IMO Guidance for the Uniform Application of Certain Rules of the 1972 Collision Regulations.

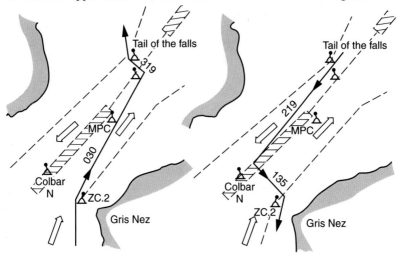

It is also possible that a vessel may only want to cross one traffic lane and the separation zone or line to join the other traffic lane at as small an angle to the general direction of traffic flow as practicable. Such a manœuvre must be carried out with caution and full awareness of the traffic moving in the scheme. The requirement to cross at right angles is not limited to a vessel crossing both lanes of a traffic separation scheme.

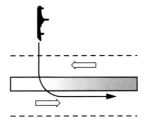

Century Dawn-Asian Energy

It was submitted by Mr Teare that on the true construction of Rule 10(c) the obligation to cross a separation lane as nearly as practicable at a right angle only applied to vessels crossing both lanes and not to vessels crossing one lane with a view to joining the other. It was further submitted that under Rule 10(b)(iii) it was the obligation of Century Dawn *to join the east bound lane at as small an angle as possible so that she was justified in crossing the westbound lane at less than a right angle. I do not accept those submissions. In my judgment, particularly in the light of the advice of the assessors which I have set out above, the obligation in Rule 10(c) applies to vessels crossing any traffic lane whether the purpose of crossing it is to cross the next lane or to join it. Of course the obligation to cross at a right angle is qualified by the expression 'as nearly as practicable'. Moreover, no attempt should be made to cross either lane in a traffic separation scheme unless it is safe to do so. (Mr Justice Clarke, 1994)*

Inshore traffic zones

Inshore traffic zones have been established alongside some traffic separation schemes with the intention of keeping coastal shipping away from traffic passing through the adjacent traffic lanes. Such zones may be relatively narrow and could become dangerous if extensively used by traffic proceeding in opposite directions.

A vessel proceeding en route to or from a port, offshore installation or structure, pilot station or any other place situated within the inshore traffic zone is permitted to use the zone. An inshore traffic zone may also be used by vessels less than 20 metres in length, sailing vessels and vessels engaged in fishing.

The use of traffic separation schemes is, for most schemes adopted by IMO, not mandatory (see also page 60 *Vessels not using the scheme*). Vessels are, of course, permitted to proceed in either direction in open water outside the scheme as an alternative to passing through the traffic lanes provided the outer limit is avoided by a wide margin in accordance with Rule 10(h).

In straits or channels where inshore zones have been established on both sides of a traffic separation scheme so that through traffic must either use the scheme or pass through the inshore zones, power-driven vessels of over 20 metres in length, not calling at any place within the zones, must pass through the appropriate traffic lane unless there are special circumstances, such as severe weather conditions or some emergency, which make it necessary to enter an inshore zone to avoid immediate danger.

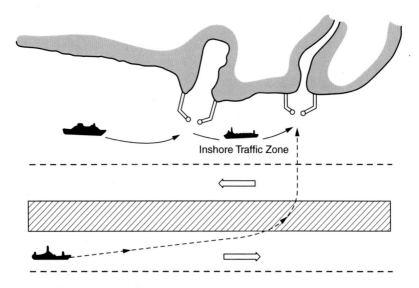

In 1989 the sixteenth Assembly of IMO approved an amendment to Rule 10(d) to clarify the circumstances in which inshore traffic zones may be used.

Separation zones

In many traffic separation schemes separation zones have been established between the lanes to separate traffic proceeding in opposite directions. A separation zone can also be established between a traffic lane and an inshore zone. These zones may only be used by vessels crossing the area covered by the separation scheme, by vessels joining or leaving a lane, by vessels engaged in fishing and by vessels obliged to enter in cases of emergency to avoid immediate danger.

The first sentence of paragraph (e) was amended in 1981 to incorporate vessels joining or leaving a lane.

A vessel crossing a separation zone which is also crossing the traffic lanes should cross as nearly as practicable at right angles to the general direction of traffic flow. However, a vessel crossing a zone in the process of joining or leaving a lane from either side must comply with paragraph (b)(iii).

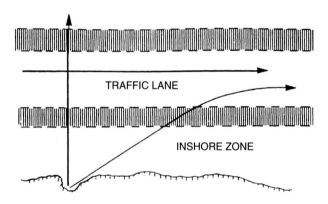

The right of a vessel to engage in fishing within a traffic separation zone is implicitly established by paragraph (e). As there is no general direction of flow within a zone vessels engaged in fishing may move in any direction but they should take account of the general principles of traffic separation schemes and refrain from proceeding in the opposite direction to the general direction of flow in the adjacent traffic lane when fishing near a lane boundary.

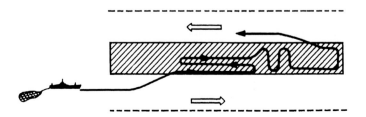

Good seamanship requires that vessels fishing within a traffic separation zone should pay particular attention to their position and to the movement of other vessels. The nets of a vessel fishing within a separation zone must not be allowed to extend across a lane in such a way as to impede the passage of vessels following the lane; see paragraph (i).

Lane terminations

A primary objective of traffic separation is to reduce the number of meeting or fine crossing situations which have been found to be particularly dangerous because of the high speed of approach. The establishment of traffic lanes has reduced the risk of collision within the lanes but has probably increased the incidence of fine crossing situations near the terminations due to converging and diverging traffic. Caution is therefore necessary when navigating in areas near the terminations at all times but this applies especially when the visibility is restricted.

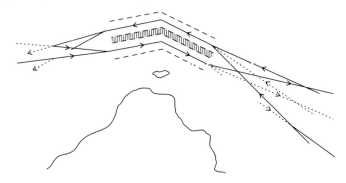

Anchoring to be avoided

One of the aims of traffic separation is to reduce the speed at which vessels approach one another by causing traffic to move along the lanes in the same direction. In a traffic stream a vessel at anchor, or a vessel underway and stopped, is therefore a source of danger, particularly in restricted visibility. The direction of the wind or stream may cause an anchored vessel to lie at a broad angle to the traffic flow which may result in her being a serious obstruction to traffic in a narrow traffic lane.

A vessel is also required to avoid anchoring in a traffic separation zone and in areas near the termination of a lane.

Vessels not using the scheme

Paragraph (h) is intended to apply mainly to vessels proceeding through the area outside the boundaries specified in the scheme in a direction opposite to the general direction of flow within the adjacent lane. The danger of vessels proceeding in opposite directions meeting one another near the outer limits of a lane was referred to on page 55. Paragraph (c) permits vessels to cross at right angles if it is not practicable to avoid crossing the lanes.

The use of some traffic separation schemes may be mandatory for all ships, certain categories of ships or ships carrying certain cargoes. If there is a mandatory requirement to use a traffic separation scheme Rule 10(h) will not apply.

Vessels engaged in fishing

A vessel is permitted to engage in fishing in traffic separation zones or traffic lanes provided she does not impede the passage of a vessel following a lane and does not proceed against the general direction of flow when fishing within a lane.

The question of whether the Rule might be interpreted as permitting a vessel engaged in fishing to proceed in a traffic lane in the opposite direction to the direction of flow was considered at the 1972 Conference. The Conference was categorically of the opinion that no vessel should be allowed to proceed against the direction of the established flow of traffic in a traffic lane. Opinion was unanimous on this point.

The following statements made in Committee II of the 1972 Conference reflect the general views expressed:

Since fishing vessels would wish to fish where the fish were, they should be allowed to do so in traffic lanes, provided they were moving in the direction of the traffic flow. (Captain B. Repkin, USSR, Chairman)

Clearly it would be impracticable to forbid fishing vessels from fishing inside traffic separation schemes. The point made about vessels sailing in the wrong direction was surely not relevant since any court would find that such a vessel had been in breach of a Rule. (Captain A. Manson, UK)

A vessel engaged in fishing outside the outer limits of the area covered by a traffic separation scheme must not allow her nets to extend into a traffic lane in such a way as to impede the passage of a vessel following the lane.

The requirement not to impede the passage of a vessel following a lane must be applied in conformity with Rule 8(f) (see pages 43–45).

Rule 10(i) and (j)

A vessel which is required not to impede the passage or safe passage of another vessel, in accordance with Rule 10(e) and (j) must also comply with the requirements of Rule 8(f) and must, when required by the circumstances of the case, take early action to allow sufficient sea room for the safe passage of the vessel whose passage is not to be impeded (see pages 43–45).

Small vessels and sailing vessels

The requirement of paragraph (j) is similar to that of paragraph 9(b) relating to narrow channels, but in traffic lanes small vessels and sailing vessels must avoid impeding the safe passage of any power-driven vessel following the lane. For this requirement the application of Rule 8(f) is relevant (see pages 43–45). Small vessels and sailing vessels are not required to avoid impeding the safe passage of power-driven vessels crossing a lane or moving against the direction of flow.

A sailing vessel, or small power-driven vessel, should, preferably, wait for a suitable opportunity to cross a traffic lane, but a power-driven vessel following a lane is not relieved of her obligation to keep out of the way if there is risk of collision with a sailing vessel.

Special signal

The International Code two letter signal 'YG' has the meaning 'You appear not to be complying with the traffic separation scheme'. The master of any vessel receiving the signal by whatever means should take immediate steps to check his course and position and any further action which may be appropriate in the circumstances.

Deep water routes

A deep water route is defined in the IMO publication *Ships' Routeing* as a route in a designated area within definite limits which has been accurately surveyed for clearance of sea bottom and submerged obstacles as indicated on the chart. It is primarily intended for use by ships which because of their draught in relation to the available depth of water are restricted in their choice of route. Through traffic not restricted by draught considerations should, if practicable, avoid following deep water routes.

A deep water route may form part of a traffic lane and be intended for use by deep draught vessels moving in the general direction of flow. The provisions of Rule 10 would apply in this case as the route would be covered by a traffic separation scheme. Deep water routes have been established which are not part of a traffic separation scheme adopted by the Organization. Rule 10 does not apply to such routes but it would, nevertheless, be prudent for vessels which cannot avoid crossing them to do so at right angles.

Deep water routes which do not form part of a separation scheme may be intended for use by one-way or two-way traffic, as indicated by arrows on the chart. Vessels using a deep water route for two-way traffic should keep to the starboard side of the route.

Cable work and safety operations

Vessels engaged in laying, servicing or picking up a submarine cable or navigation mark and vessels engaged in surveying are included in the categories of vessels to be regarded as being restricted in their ability to manœuvre, as prescribed in Rule 3(g). Vessels engaged in the above activities are, therefore, likely to be privileged with respect to other vessels but they would normally be expected to comply with the provisions of Rule 10 and to either cross traffic lanes at right angles or proceed along them in the general direction of traffic flow.

It has been found that strict compliance with Rule 10 would make it difficult, if not impossible, to effectively carry out essential operations such as hydrographic surveying and the servicing of cables. The IMO General Assembly therefore adopted, in 1981, paragraphs (k) and (l) which exempt a vessel engaged in an operation for the maintenance of safety of navigation, or in the laying, servicing or picking up of a submarine cable, within a traffic separation scheme from complying with Rule 10 to the extent necessary to carry out the operation.

Vessels engaged in the operations referred to in paragraphs (k) and (l), and which are exhibiting the lights or shapes prescribed in Rule 27(b), may thus be exempted from compliance with Rule 10 and may even be justified in proceeding in a direction opposite to the general direction of traffic flow in a traffic lane. However, they are expected to comply with Rule 10 whenever possible.

The Government, or appropriate authority, must be notified of, and must authorise, such operations. Information about vessels working in a traffic separation scheme shall, so far as practicable, be promulgated beforehand by Notices to Mariners and by subsequent radio warnings before, and at regular intervals during, the operations.

In the General Provisions on Ships' Routeing it is specified that such operations shall as far as possible be avoided in conditions of restricted visibility.

Precautionary area

A precautionary area is defined in the IMO principles of ships' routeing as a routeing measure comprising an area within definite limits where ships must navigate with particular caution and within which the directions of traffic flow may be recommended.

Section II – Conduct of Vessels in Sight of One Another

RULE 11

Application

Rules in this Section apply to vessels in sight of one another.

COMMENT:

Rule 3(k) states that vessels shall be deemed to be in sight of one another only when one can be observed visually from the other. The Rules in Section II do not apply to a vessel which has detected another vessel by radar, and has established that risk of collision exists, if the other vessel cannot be sighted visually. Rule 19 of Section III applies only to vessels navigating in or near an area of restricted visibility which are not in visual sight of one another. In restricted visibility, therefore, vessels may initially have to comply with Rule 19 of Section III then subsequently have to comply with the Rules of Section II when they come into visual sight of one another. A vessel is unlikely to be excused for not complying with the appropriate Rule of Section II if it is considered that failure to sight the other vessel was due to a bad visual look-out.

It is conceivable that instantaneous sighting may not occur, even if both vessels are keeping an efficient visual look-out, due to such factors as different intensities of navigation lights or to patches of low fog obscuring the bridge of one vessel but not her masthead lights. A vessel must comply with the Rule which relates to the situation which applies at the particular instant.

In the discussions which took place before the 1972 Conference serious consideration was given to the possibility of formulating one set of manœuvring rules which would apply in all conditions of visibility. The Conference decided against adopting this principle, however, mainly because it is usually possible for vessels to sight one another in sufficient time to recognise the lights or shapes being displayed so that the degree of responsibility can be based on the vessel's ability to take effective avoiding action.

Vessels engaged in such activities as fishing or underwater operations and vessels not under command may be incapable of manœuvring effectively. A slow vessel being overtaken by a vessel of high speed may not observe the overtaking vessel until it is too late to get clear by her own action. Even in a crossing situation involving two power-driven vessels, if both were to be required to keep out of the way, the vessel expected to pass ahead of the other is likely to be less capable of achieving a safe passing distance by her own action than the vessel expected to cross astern of the other.

Rules 13 and 18 of Section II are based on the principle of allocating prime responsibility to the vessel which will usually be more capable of keeping out of the way. If no such distinction were made the vessel with the greater ability to take effective avoiding action would be more likely to wait for the other to keep out of the way.

It is, of course, not possible to allocate greater responsibility to the vessel which is more capable of taking avoiding action when the visibility is restricted and the vessels are not in visual sight of one another as all vessels do not have an operational radar and a means of identifying a hampered vessel by radar has not yet been established. Fortunately, there are few areas of the world in which serious restriction of visibility is likely to be frequently experienced.

The 1972 Conference did adopt some changes which resulted in greater compatibility between the Rules for vessels in visual sight of one another and the Rule for restricted visibility. The most important of these is that a privileged vessel is now permitted to act at an earlier stage when vessels are in sight of each other. In restricted visibility more emphasis has been placed on the avoidance of an alteration of course to port for a vessel detected forward of the beam, which is in accordance with the principles established in Rules 14, 15 and 17(c) of Section II.

RULE 12

Sailing vessels

(a) **When two sailing vessels are approaching one another, so as to involve risk of collision, one of them shall keep out of the way of the other as follows:**
   (i) **when each has the wind on a different side, the vessel which has the wind on the port side shall keep out of the way of the other;**
   (ii) **when both have the wind on the same side, the vessel which is to windward shall keep out of the way of the vessel which is to leeward;**
   (iii) **If a vessel with the wind on the port side sees a vessel to windward and cannot determine with certainty whether the other vessel has the wind on the port or on the starboard side, she shall keep out of the way of the other.**
(b) **For the purposes of this Rule the windward side shall be deemed to be the side opposite to that on which the mainsail is carried or, in the case of a square-rigged vessel, the side opposite to that on which the largest fore-and-aft sail is carried.**

COMMENT:

Rules 8, 13, 16 and 17(a), (b) and (d) also apply when two sailing vessels are approaching one another so as to involve risk of collision. A sailing vessel must take early and substantial action to achieve a safe passing distance. The other vessel must initially try to keep her course and speed, but wind changes may make this difficult.

Rule 12 will not apply if one of the vessels under sail is also using propelling machinery as such a vessel is considered to be a power-driven vessel.

Exceptions

A sailing vessel approaching another sailing vessel from a direction more than 22.5° abaft her beam is an overtaking vessel and must keep clear regardless of wind direction, as Rule 13 over-rides Rule 12. A sailing vessel must also keep out of the way of any sailing vessel which is engaged in fishing, or not under command, and showing the appropriate lights or shapes.

Ambiguous case

Paragraph (a)(iii) covers the ambiguous case in which a sailing vessel with the wind on the port side sees the green sidelight of another sailing vessel to windward at night and is unable to determine whether the other vessel has the wind on the same side and is required to keep out of the way or has the wind on the starboard side and is required to keep her course and speed. In such circumstances she is required to keep clear, preferably by bearing away, taking account of the possibility that the other vessel may take avoiding action.

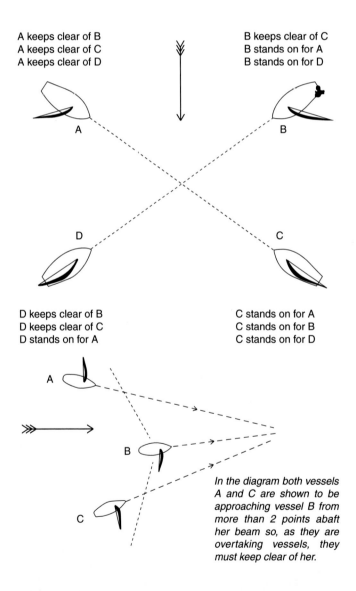

A keeps clear of B
A keeps clear of C
A keeps clear of D

B keeps clear of C
B stands on for A
B stands on for D

D keeps clear of B
D keeps clear of C
D stands on for A

C stands on for A
C stands on for B
C stands on for D

In the diagram both vessels A and C are shown to be approaching vessel B from more than 2 points abaft her beam so, as they are overtaking vessels, they must keep clear of her.

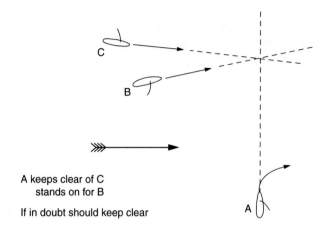

A keeps clear of C
stands on for B

If in doubt should keep clear

A sailing vessel which has the wind on the starboard side and sees the red sidelight of another sailing vessel to windward at night may also be unable to determine whether the other vessel has the wind on the port side or the starboard side. However, the other vessel is required to keep out of the way in either case so course and speed should be maintained.

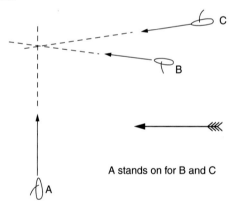

A stands on for B and C

RULE 13

Overtaking

(a) Notwithstanding anything contained in the Rules of Part B, Sections I and II, any vessel overtaking any other shall keep out of the way of the vessel being overtaken.

(b) A vessel shall be deemed to be overtaking when coming up with another vessel from a direction more than 22.5 degrees abaft her beam, that is, in such a position with reference to the vessel she is overtaking, that at night she would be able to see only the sternlight of that vessel but neither of her sidelights.

(c) **When a vessel is in any doubt as to whether she is overtaking another, she shall assume that this is the case and act accordingly.**

(d) **Any subsequent alteration of the bearing between the two vessels shall not make the overtaking vessel a crossing vessel within the meaning of these Rules or relieve her of the duty of keeping clear of the overtaken vessel until she is finally past and clear.**

COMMENT:

Paragraph (a) of this Rule was amended in 1981, the words 'of this section' in the first line being replaced by the words 'of Part B Sections I and II'.

The amendment was made to make it clear that a vessel proceeding along a narrow channel or traffic lane is expected to keep out of the way of any vessel she is overtaking, including a sailing vessel, a small power-driven vessel and a vessel engaged in fishing. Rule 13 takes precedence over Rules 12 and 18 so that a sailing vessel overtaking another sailing vessel must keep out of the way, irrespective of wind direction, and a vessel from any of the categories listed in Rule 18 must keep out of the way of any vessel which she is overtaking. Under the 1960 Regulations there was some doubt as to whether a vessel not under command or a vessel which is now considered to be 'restricted in her ability to manœuvre' was required to keep clear of a vessel which she was overtaking.

A vessel which is overtaking another vessel will usually have little difficulty in keeping out of the way, by either helm action or engine action, as there is unlikely to be a high speed of approach. A hampered vessel which is unable to make a substantial alteration of course will normally be able to avoid collision by reducing her speed. Prime responsibility for keeping out of the way is allocated to the overtaking vessel in every case as that vessel must be proceeding at greater speed and is more likely to sight the vessel being overtaken at an early stage.

Subsequent alterations of bearing

Every vessel overtaking any other is obliged to keep clear of the overtaken vessel. This Rule applies even to cases in which the bearing is changing appreciably. If a vessel coming up relatively close to another vessel from any direction more than 22.5° abaft her starboard beam, draws ahead, and then subsequently turns to port to come on to a crossing course, she is not relieved of the duty of keeping clear. However, if she is a considerable distance away from the overtaken vessel so that there is no risk of collision when she passes her the Rules would not apply at that time, and the other vessel would be obliged to keep clear in a subsequent crossing situation bringing risk of collision between the same two vessels.

In the upper diagram the overtaking vessel is shown to be at a relatively close distance when she first comes to within 22.5° abaft the beam of the other vessel. The overtaking vessel has the obligation to keep out of the way until she is finally past and clear.

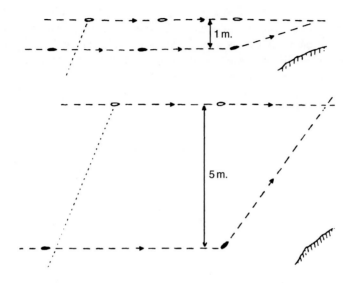

The lower diagram shows the faster vessel initially approaching from more than 22.5° abaft the beam of the other at a relatively large distance (over 5 miles) so that, although the vessels are likely to be in sight of one another, risk of collision could hardly be considered to apply as they are on parallel courses. In this case the slower vessel will be required to keep out of the way if the faster vessel turns on to a converging course which brings risk of collision.

In doubtful cases the faster vessel should assume the obligation to keep out of the way if it becomes necessary to turn onto a crossing course and risk of collision is found to exist.

Baines Hawkins–Moliere

It appears to me that the view taken by the Counsel for the Plaintiffs with regard to the obligation of an overtaking vessel is, on the whole, a sound one; that is to say, that when a vessel is an overtaking vessel within the strict sense of the word, that is, a vessel which is within the area lighted by the stern light, and then comes, while she is still advancing into a position in which she sees a side-light, sometimes, if not always, her obligation as an overtaking vessel to keep out of the way of the other still continues. It is admitted by the Counsel for Defendants that would be so if, at the time of her seeing a side-light, there was risk of collision. I do not see how any other admission than that could be made, because it would be strange indeed, if a vessel overtaking came in sight of one of the side-lights, and then suddenly, when there was risk of a collision, threw on the other the obligation of keeping out of the way. It may, on the other hand, be that, when there is no risk of collision at the time – if, for example, the vessel comes within sight of a side-light at a considerable distance – the crossing rule may come into force; but, in this case, I am satisfied that the facts are such that one cannot suppose that the obligation of the Moliere, as an overtaking vessel, was over. (Sir F. Jeune, 1893)

In the case of *Auriga–Manuel Campos,* 1976, Mr Justice Brandon held that risk of collision did not exist when the *Auriga,* proceeding at 15 knots, was bearing more than 22.5° abaft the beam of the *Manuel Campos,* proceeding at 12½ knots, as the courses were diverging by 7° and the vessels were shaping to pass abeam at about 3 miles. The *Auriga* altered course about 30° to port, in the process of navigation, when only a few degrees abaft the starboard beam of the *Manuel Campos.* It was held that the Crossing Rule applied but the *Auriga* was held to be 60% to blame for bad look-out and, in particular, for setting a converging course which created a dangerous situation.

In the case of *Olympian–Nowy Sacz* it was held by the Court of Appeal, 1977, that the Overtaking Rule (previously Rule 24) begins to operate when a vessel is coming up with another from more than 22½° abaft the beam and may apply before there is risk of collision. Sir David Cairns said:

... If, therefore, ships came in sight of each other when many miles apart, we think it would be wrong, whatever their relative positions and courses may have been, to say that one was 'coming up with' the other. It does not, however, follow that for one to be coming up with the other there must be risk of collision between them. For instance, if two ships are on parallel courses and one is abaft the other and travelling faster, we think a time would come when the faster ship should be considered to be coming up with the other, provided that the courses were not more than a few cables apart, even though if each ship maintained its course there would be no risk of collision.... We would hold accordingly that Rule ... begins to operate before there is risk of collision and as soon as it can properly be said that the overtaking ship is coming up with the overtaken ship. When exactly that will be may not always be easy to determine but we see no reason to suppose that it will be any more difficult than the decision as to when the situation involves a risk of collision.

In the *Manchester Regiment–Clan Mackenzie,* the two vessels were proceeding in approximately the same direction when the leading vessel, which had the other about 22.5° on her starboard quarter, altered course eight points to starboard in the process of adjusting compasses. It was held that up to the time of the alteration the Regulations did not apply (the vessels were distant about 2 miles) and the vessels were considered to be crossing prior to the collision.

The distance at which the Rules apply will depend largely on the speed of approach; it may be less than a mile in the case of two slow vessels proceeding on similar courses with little difference in speed.

Action to be taken by the overtaking vessel

A vessel which is overtaking another vessel is required to keep out of the way and to pass at a safe distance. She is not required to avoid crossing ahead of the other vessel but altering course, or reducing speed, in order to pass astern of the vessel being overtaken may be the safest form of avoiding action. The overtaking vessel is also required to take action at an early stage. If action is not taken in good time there is a danger that the vessel being overtaken may take action which could confuse the situation.

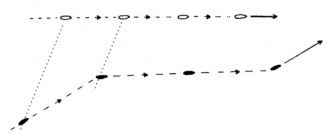

A power-driven vessel which approaches another power-driven vessel from a direction approximately 22.5° abaft her beam may be in doubt as to whether she is an overtaking vessel or a crossing vessel. There should not be any doubt at night because a crossing situation is indicated if a side-light is seen, but the aspect cannot be determined accurately by day. Rule 13(c) requires such a vessel to assume that she is overtaking and keep out of the way. As the other vessel may ascertain that a crossing situation exists, and take action to avoid a vessel crossing from her own starboard side, the vessel which is to starboard should preferably turn on to a parallel course and subsequently pass ahead.

Interaction

It is now generally accepted, as a result of model tests and practical experience gained by ships replenishing at sea, that when two ships pass close to one another, on roughly parallel courses, forces of attraction and repulsion are set up between them. This effect is known as interaction. It will be greatest in shallow water and when the two vessels are moving at high speed in the same direction with little difference of speed between them. In the case of two vessels passing on opposite courses interaction will have little effect, but in overtaking situations the course of one or both of the vessels may be affected to an appreciable extent, especially when a large vessel is overtaking a smaller one.

The maximum distance between two vessels at which interaction may be noticed will vary with the size and speed of the ships and the depth of water. It may be over 300 metres in some cases. Even in deep water interaction may be experienced by fast vessels overtaking at close distances. The *Queen Mary–Curacao* collision was considered to have been caused partly by interaction yet the depth of water in the area was about 120 metres.

Overtaking vessels should not attempt to pass too close in open waters when there is plenty of room to manœuvre. In narrow channels it may well be dangerous to overtake another vessel which is itself moving at high speed.

Queen Mary–Curacao

No doubt the effect of the forces of interaction are very imperfectly known, and one cannot impute to the captains of the two ships any expert or exact knowledge of them, but I should have expected some allowance to have been made for their coming into play, in the sense that the ships should not have been allowed to approach so near to one another as to run a risk of their coming into action. (Lord Porter, 1949, House of Lords)

When a ship is moving at any appreciable speed there is a region of increased pressure in the water near the bow and stern and a region of decreased pressure amidships. If two ships pass close to one another on parallel courses forces of attraction and repulsion may be experienced between them. The following diagrams indicate the possible effects.

As the stern of vessel A overtakes the stern of vessel B there will be a repulsive force between them so that there will be a tendency for vessel B to swing her bows across the path of vessel A (fig. 1). The *Queen Mary–Curacao* collision has been attributed to this effect.

Later the turning moment is reversed, and as the bows of the two ships draw level vessel B will tend to swing outward as shown in fig. 2.

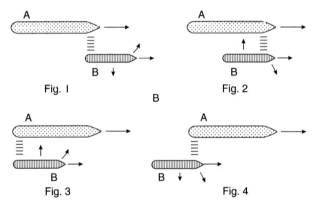

Fig. I

Fig. 2

Fig. 3

Fig. 4

When the sterns of the two vessels come together there will be a repulsive force between them so that once again there will be a tendency for the bows of vessel B to swing inwards (fig. 3). The *Olympic–Hawke* collision may have been caused by this effect.

Finally as the stern of vessel A passes the stern of vessel B the turning moment on vessel B will again be reversed (fig. 4).

Overtaking in narrow channels and traffic lanes

Rule 13 overrides other Rules in Sections I and II but does not give small vessels, sailing vessels and vessels engaged in fishing the right to impede the passage of any vessel overtaking them when they are within a narrow channel or following a traffic lane. Small power-driven vessels and sailing vessels should keep clear of the deeper part of a narrow channel on the approach of any vessel which may be unable to navigate outside the channel. Vessels engaged in fishing must not impede the passage of any vessel navigating within a narrow channel or following a traffic lane.

A vessel following a traffic lane, or proceeding along a narrow channel to which it is restricted, when overtaking any vessel, including a small vessel, sailing vessel or vessel engaged in fishing should reduce speed if necessary, or take whatever avoiding action she can that is safe and practicable. When vessels are in sight of one another and there is risk of collision the prime responsibility for keeping out of the way rests with the overtaking vessel. When, in a narrow channel, overtaking can take place only if the vessel to be overtaken has to take action to permit safe passing, Rule 9(e)(ii) effectively requires the overtaking vessel to keep out of the way whether or not the other vessel indicates agreement and takes appropriate action.

The procedure to be adopted when overtaking can only take place by mutual agreement in a narrow channel is described in Rule 9(e)(i) (see page 49). The effects of interaction, bow cushion and bank suction must be taken into account when overtaking in a narrow channel.

In the case of the *Ore Chief–Olympic Torch,* 1974, Mr Justice Brandon asked the Assessors what risks should a prudent pilot have realised were involved in overtaking at a particular part of the River Schelde. Their answer was as follows:

a. *collision due to close proximity of the vessels;*
b. *interaction between the vessels causing one to sheer towards the other or towards the bank and leading to collision or grounding;*
c. *interaction between either vessel and the bank causing her to sheer towards or away from the bank, again leading to collision or grounding.*

The judge accepted this advice and found the *Ore Chief* negligent in overtaking where she did.

RULE 14

Head-on situation

(a) **When two power-driven vessels are meeting on reciprocal or nearly reciprocal courses so as to involve risk of collision each shall alter her course to starboard so that each shall pass on the port side of the other.**

(b) **Such a situation shall be deemed to exist when a vessel sees the other ahead or nearly ahead and by night she could see the masthead lights of the other in a line or nearly in a line and/or both sidelights and by day she observes the corresponding aspect of the other vessel.**

(c) **When a vessel is in any doubt as to whether such a situation exists she shall assume that it does exist and act accordingly.**

COMMENT:

Rule 14 differs from the other Rules in Section II in placing equal responsibility for keeping out of the way on each of the two vessels involved and in stating, specifically, what action should be taken by each vessel. It applies only to power-driven vessels.

Application

Rule 14 applies when two power-driven vessels are meeting on reciprocal or nearly reciprocal courses so as to involve risk of collision. A dangerous situation may arise if the two vessels appear likely to pass at close distance starboard to starboard. A requirement to avoid crossing ahead only applies on crossing courses (Rule 15) so vessels meeting starboard to starboard so as to involve risk of collision should make an early and substantial alteration of course to starboard to achieve a port to port passing. In such circumstances it may be dangerous to turn to starboard at close range.

In the case of *Sea Star-Horta Barbosa*, 1976, the two vessels were found to have been on reciprocal courses so as to pass at close distance starboard to starboard. At a late stage the *Sea Star* made a substantial alteration of course to starboard. The *Sea Star* sank with loss of life, *and* was found mainly to blame for attempting to cross ahead at close range.

Rule 14 is apparently not intended to apply to cases in which, from a vessel ahead or nearly ahead, one sidelight can be seen but the other obscured.

In order to avoid possible dark lanes immediately ahead of a ship the sidelights are screened so as to show approximately 2° across the bow (see page 107). The effect of yawing must also be taken into account; this will vary with the steering arrangements and steering qualities of the ship.

The wording of the Rule makes it clear that it is the direction of the ship's head, and not the course made good, which must be used to determine whether vessels are meeting end on or crossing. This may be important in conditions of strong wind or tide, where one vessel is drifting more rapidly than another, so that one vessel may see another end on fine on the bow, and the bearing may remain constant.

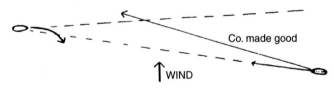

If one vessel sees the other end on the starboard bow, as shown in the figure, both vessels may be tempted to alter course to port. Such action would not be in accordance with the general principles of the Rules. The vessel with the other on her own starboard side is required to keep out of the way by Rule 15 and should preferably alter course to starboard to avoid crossing ahead of the other vessel, with respect to her course made good. The vessel which sees the other ahead should initially maintain course and speed, but may take action if the give-way vessel fails to keep clear at an early stage. A substantial alteration of course to starboard would again be the best form of avoiding action.

In the case of *British Engineer–Karanan*, 1945, the *British Engineer* was blamed for altering course to port for a green light approximately ahead. The *Karanan* altered course to starboard. Both vessels were being affected by a strong tide setting across the approaches to Belfast Lough.

Paragraph (c) makes it clear that when a vessel is in doubt as to whether a meeting or crossing situation exists, or is in doubt as to whether the approaching ship is an ordinary power-driven vessel or a hampered vessel, she should assume that Rule 14 applies and alter course to starboard. The fact that a stand-on vessel is permitted to take action before getting so close that collision cannot be avoided by the give-way vessel alone, provided that, if power-driven, she does not turn to port for a vessel crossing from her own port side, makes it less essential to draw a clear distinction between meeting and crossing situations.

Whether power-driven vessels are meeting on reciprocal courses or crossing at a fine angle it is important that neither vessel should alter course to port. If it is thought necessary to increase the distance of passing starboard to starboard this implies that there is risk of collision. Several collisions have been caused as a result of one vessel altering course to port to increase the passing distance and the other vessel turning to starboard.

RULE 15

Crossing situation

When two power-driven vessels are crossing so as to involve risk of collision, the vessel which has the other on her own starboard side shall keep out of the way and shall, if the circumstances of the case admit, avoid crossing ahead of the other vessel.

COMMENT:

Coastal waters

Rule 15 will normally apply to power-driven vessels crossing in coastal waters, in the process of rounding buoys or headlands, but in the case of the *Alcoa Rambler–Norefjord,* 1949, it was held that the Crossing Rule did not apply as the two vessels had collided in a congested area and the stand-on vessel had been constantly changing her course.

Crossing in narrow channels and traffic lanes

Power-driven vessels in a crossing situation in a narrow channel or traffic lane must normally comply with Rule 15 but all vessels are required to avoid crossing a narrow channel if such crossing impedes the passage of a vessel which can safely navigate only within the channel (Rule 9(d)) and a power-driven vessel of less than 20 metres in length must not impede the safe passage of a power-driven vessel following a traffic lane (Rule 10(j)).

Although vessels proceeding in opposite directions in a bending narrow channel may come into a crossing situation Rules 15 and 17 do not apply. Each vessel must comply with Rule 9(a) and keep as near to the outer limit which lies on her starboard side as is safe and practicable.

Empire Brent–Stormont

As I understand the principles which apply in narrow channels, it has been laid down for many, many years that although the crossing rule does from time to time have to be applied in narrow channels (when, for instance, a vessel which is crossing the channel has to act in relation to a vessel which is proceeding up or down the channel), nevertheless, when vessels are approaching each other, navigating respectively up and down the channel, it is [Article 25] of the Collision Regulations which applies and applies exclusively. There is no room in such a situation for applying the provisions of the crossing rule at the same time as the provisions of the narrow channel rule, because the requirements under the rules are different. I have no hesitation in saying that as between a vessel coming up and a vessel going down, approaching each other in that way in a narrow channel like the Mersey, the narrow channel rule, and the narrow channel rule only, is the rule which has to be applied. (Mr Justice Willmer, 1948)

Hampered vessels

Rule 15 does not apply to two power-driven vessels crossing so as to involve risk of collision if one of the power-driven vessels is not under command, restricted in her ability to manoeuvre or engaged in fishing. Rule 18 applies in such circumstances. An ordinary power-driven vessel which encounters a vessel in one of the above categories, crossing so as to involve risk of collision from her own port side, is required to keep out of the way, but is not required to avoid crossing ahead. An alteration of course to starboard may be the best form of avoiding action if there is any doubt as to whether the other vessel is actually hampered due to the difficulty of recognising the day signal or lights.

A vessel which is engaged in a towing operation such as severely restricts the towing vessel and her tow in their ability to deviate from their course is 'restricted in her ability to manoeuvre' and is not required to keep out of the way of a power-driven vessel crossing from her starboard side, provided she is displaying the lights or shapes prescribed in Rule 27(b). No special privilege is granted to other vessels engaged in towing which are to be considered as ordinary power-driven vessels for the purpose of the Steering and Sailing Rules. The extra lights prescribed for the towing vessel are intended to indicate the extra length of obstruction to be expected and to give warning that there is a towing line stretching between the two vessels. A power-driven vessel which sees a tow on the port bow crossing so as to involve risk of collision should also take account of possible limitations of manoeuvring ability in considering when avoiding action is permitted by Rule 17(a)(ii) or required by Rule 17(b).

A vessel which is constrained by her draught is permitted to exhibit the lights or shapes prescribed in Rule 28 to indicate her limited manoeuvrability but is not relieved of her obligation to comply with the other Rules of this Section as a power-driven vessel (see pages 90–91). A vessel constrained by her draught is, therefore, expected to keep out of the way of a power-driven vessel which is crossing from her starboard side so as to involve risk of collision. Other vessels should take full account of the limited manoeuvrability of a vessel constrained by her draught in considering whether to take early action to allow sufficient sea room for safe passage in accordance with Rule 8(f) (see page 43) or when to take action in accordance with Rule 17 (see page 80).

Avoid crossing ahead

The requirement to avoid crossing ahead only applies in a crossing situation in which there is risk of collision. It does not apply at long ranges, before risk of collision begins to apply, or to cases in which the bearing is appreciably changing. If there is a possibility of risk of collision the give-way vessel must avoid crossing ahead.

King Stephen–Ashton

The only way in which the defendants can escape from liability, if these Rules apply, is by showing that the vessels were not crossing so as to involve risk of collision.... The question I have to consider is whether it can be said that at the outset there was no risk of collision. The ground upon which the defendants put it must be that having regard to their speed, and the other vessel approaching them at a slower speed on their starboard side, there really was no risk of collision. That is a view that neither I nor the Elder Brethren can accept, because, although it is said that the vessel broadened, she broadened very slightly.... The defendants were within those Rules, and their vessel ought not to have attempted to cross ahead of the other ship. (Sir Gorell Barnes, 1905)

In a crossing situation a power-driven vessel is required to avoid crossing ahead of a power-driven vessel on her own starboard side, if there is risk of collision, but is not directed to cross astern. An alteration of course to starboard will usually be the best method of keeping out of the way of a vessel which is on the starboard bow, but a reduction of speed or a substantial alteration of course to port would be preferable in order to avoid collision with a vessel approaching from near the starboard beam (see pages 38–39).

Vessel lying stopped

A power-driven vessel which is under way but stopped must, unless she is not under command, or restricted in her ability to manœuvre, keep out of the way of another vessel which approaches so as to involve risk of collision from any direction between right ahead and 22.5° abaft the beam on her starboard side. The approaching ship must not be expected to take avoiding action. A vessel lying stopped with her engines ready for manœuvre is not entitled to show any special lights or shapes to indicate that she is privileged and must comply with Rules 14, 15 and 18.

To emphasise the requirement that a vessel lying stopped should comply with the Steering and Sailing Rules as a vessel under way IMO has approved Guidance on the application of Rule 3(i) (see page 9).

In the case of *Lucania–Broomfield*, 1905, it was held in the Admiralty Division that a steam trawler lying with engines stopped, waiting for the tide, and exhibiting the masthead lights and sidelights of a vessel under way was alone to blame for failing to take steps to avoid collision with a power-driven vessel approaching from her starboard side.

RULE 16

Action by give-way vessel

Every vessel which is directed to keep out of the way of another vessel shall, so far as possible, take early and substantial action to keep well clear.

COMMENT:

The provisions of Rule 8 concerning action to avoid collision apply in any condition of visibility and must therefore be complied with by vessels in visual sight of one another. Any alteration of course or speed should be made in ample time and be large enough to be readily apparent to another vessel, action shall be such as to result in passing at a safe distance, the effectiveness of avoiding action must be checked and a give-way vessel should, if necessary, slacken her speed or take all way off.

RULE 17

Action by stand-on vessel

(a) (i) **Where one of two vessels is to keep out of the way the other shall keep her course and speed.**

    (ii) **The latter vessel may however take action to avoid collision by her manœuvre alone, as soon as it becomes apparent to her that the vessel required to keep out of the way is not taking appropriate action in compliance with these Rules.**

(b) **When, from any cause, the vessel required to keep her course and speed finds herself so close that collision cannot be avoided by the action of the give-way vessel alone, she shall take such action as will best aid to avoid collision.**

(c) **A power-driven vessel which takes action in a crossing situation in accordance with sub-paragraph (a)(ii) of this Rule to avoid collision with another power-driven vessel shall, if the circumstances of the case admit, not alter course to port for a vessel on her own port side.**

(d) **This Rule does not relieve the give-way vessel of her obligation to keep out of the way.**

COMMENT:

One of two vessels

A vessel is only required to maintain her course and speed in a two vessel situation. In the unlikely event of one vessel finding herself on a collision course with two other vessels at the same time, being in one case the give-way vessel and in the other case the stand-on vessel, she could not be expected to keep out of the way of one vessel and maintain her course and speed for the other.

One vessel is to keep out of the way

Rules 12, 13, 15 and 18 require one of two vessels to keep out of the way. The 'give-way vessel' is required to take early and substantial action to keep well clear by Rule 16. Rule 17 lays down provisions for the other vessel, referred to as the 'stand-on vessel'.

Rule 17 does not apply if the two vessels concerned are not in visual sight of each other, or if there is no risk of collision. This means that, for instance, a power-driven vessel which detects another vessel approaching from the port bow, or from more than 22.5° abaft the beam, and determines by radar that the bearing is not changing, is not required to keep her course and speed if the vessel cannot be sighted visually. There is also no obligation to keep course and speed for a vessel sighted at long range, before risk of collision begins to apply, even though the bearing may not be appreciably changing.

A United Kingdom proposal to introduce a 'Long Range Rule', which was intended to make it clear that disengagement was permitted at long range, was not accepted by the 1972 Conference. The Chairman stated that he had always assumed that a vessel had the right to take action early in an encounter to disengage from what might become a dangerous situation and this view was shared by other delegates. Court decisions have also been made to this effect. The Rules in Section II generally require one of two vessels to keep out of the way when risk of collision exists and risk of collision has not been considered to apply at long ranges (see pages 26–27).

Keep course and speed

A vessel which is required to keep her course and speed does not necessarily have to remain on the same compass course and maintain the same engine revolutions.

In the *Windsor–Roanoke,* 1908, both vessels were bearing down on the Rotterdam pilot boat, on crossing courses, when the *Roanoke,* while signalling for a pilot, stopped her engines to take the pilot on board. Although the *Roanoke* was the stand-on vessel, she was held to be justified in her manœuvre, as the other vessel should have known what she was doing. Lord Alberstone said:

In my judgment, 'course and speed' mean course and speed in following the nautical manœuvre in which, to the knowledge of the other vessel, the vessel is at the time engaged. It is not difficult to give many instances which support this view. The 'course' certainly does not mean the actual compass direction of the heading of the vessel at the time the other is sighted.... A vessel bound to keep her course and speed may be obliged to reduce her speed to avoid some danger of navigation, and the question must be in each case, 'is the manœuvre in which the vessel is engaged an ordinary and proper manœuvre in the course of navigation which will require an alteration of course and speed; ought the other vessel to be aware of the manœuvre which is being attempted to be carried out?'.

In the *Manchester Regiment–Clan Mackenzie,* 1938, both vessels were heading in the same direction at a distance of two to three miles from each other, when the one ahead, which was adjusting compasses, swung about eight points to starboard, bringing the other on to her starboard bow. It was held that the Rules were not applicable at the time of the alteration, so that the vessel adjusting compasses was the give-way vessel. With reference to the adjusting of compasses, the President, Lord Merriman, said:

In my opinion, if I were to hold that the manœuvres convenient for adjusting compasses are in the same category as the recognised nautical manœuvre of picking up a pilot, I should be tearing up the Steering and Sailing Rules without the slightest warrant.

May take action

A stand-on vessel is not specifically required to take action to avoid collision as soon as it becomes apparent that the give-way vessel is not taking appropriate action. She is permitted to keep her course and speed until collision cannot be avoided by the give-way vessel alone. However, the provision for permissive action places greater emphasis on the obligation of the stand-on vessel to continuously assess the situation when risk of collision exists to indicate any doubt by use of the signals prescribed in Rule 34(d) and, subsequently, to take action before collision becomes inevitable. A stand-on vessel which fails to take action in sufficient time to avoid collision by her own manœuvre is likely to be held at fault if a collision should occur. The difficulty of determining the precise moment when action becomes compulsory is less likely to be accepted as a valid excuse for waiting too long now that a stand-on vessel is permitted to manœuvre at an earlier stage.

Earliest moment for permitted action

When risk of collision first begins to exist the stand-on vessel must keep her course and speed. The give-way vessel is required to keep out of the way in good time and to take substantial action which will result in passing at a safe distance. The method of keeping out of the way is not specified but in the case of two power-driven vessels crossing the give-way vessel must avoid crossing ahead. A stand-on vessel which takes avoiding action before it can reasonably be assumed that the give-way vessel is not taking appropriate action is likely to be held mainly to blame if practically simultaneous action by the give-way vessel causes a confused situation which results in collision.

The stand-on vessel is required to keep her course and speed until it becomes apparent that the give-way vessel is either failing to take action in ample time or failing to take sufficient action to achieve a safe passing distance. The obligations of the give-way vessel are specified in Rules 8 and 16. Rule 16 requires every give-way vessel to take early and substantial action and the provisions of Rule 8 include requirements to take action which will be readily apparent to the other vessel and will result in passing at a safe distance.

Action should not be taken by the stand-on vessel without first determining that risk of collision does in fact exist. Compass bearings should be observed accurately and the radar should be used to measure the range of the approaching vessel.

The earliest moment for permitted action will obviously be related to the range and the rate of change of range.

In the open sea a give-way vessel which approaches to within a distance of about two miles in a crossing situation involving two merchant ships can usually be considered to have waited too long, but smaller or greater distances may apply depending upon the size and manœuvrability of the vessels and depending particularly upon the rate of approach.

Action to be taken by the stand-on vessel

When vessels are in sight of one another any vessel which fails to understand the intentions or actions of an approaching vessel, or is in doubt whether the other is taking sufficient action to avoid collision, is required by Rule 34(d) to immediately indicate such doubt by giving at least five short and rapid blasts on the whistle. The sound signal may be supplemented by a light signal of at least five short and rapid flashes which may be more effective as a 'wake-up' signal, especially at distances over 2 miles. If these signals bring no immediate response further precautionary measures should be taken aboard the stand-on vessel, depending upon the circumstances, such as calling the master, changing to manual steering and putting the engines on stand-by.

A stand-on vessel which takes permitted action to avoid collision by her manœuvre alone, when it becomes apparent that the give-way vessel is not taking appropriate action, must obviously take full account of the possibility that the give-way vessel may also take simultaneous or subsequent action. The stand-on vessel should avoid taking action which is likely to conflict with the probable action of the give-way vessel.

Rule 17(c) requires a power-driven vessel to avoid turning to port to avoid collision with another power-driven vessel crossing from her own port side. In such a situation the give-way vessel is required to avoid crossing ahead and is likely to turn to starboard. An alteration of course to port may also be dangerous for any stand-on vessel, including a hampered vessel, sailing vessel or vessel being overtaken, which has a give-way vessel approaching from the port side.

Rule 8(e) requires a vessel to slacken her speed if necessary to avoid collision. A reduction of speed made by the stand-on vessel would make it more difficult for the give-way vessel to cross astern, which is her most likely method of keeping out of the way. An increase of speed might even be appropriate in certain circumstances, particularly in association with helm action, but any alteration of speed should be substantial and a vessel is unlikely to be proceeding at reduced speed if the Rules of Section II apply. A change of speed is usually slow to take effect and will be less readily apparent to the other vessel than helm action.

An alteration of course away from the direction of the other vessel will usually be the safest manœuvre, if it is made in sufficient time. Such a manœuvre could hardly contribute to a collision, even if made too early, provided it has been established that the bearing is not, in fact, closing on the bow. Turning away from the other vessel in a crossing situation will, at least, slow down the rate of approach. If the give-way vessel is approaching from less than about 60° on the bow the best action may be to turn away until the other vessel is approximately abeam, but if the give-way vessel is overtaking or approaching from near the beam an alteration on to a parallel or slightly diverging course would probably be the safest action (see also pages 168–170).

Although turning away from the give-way vessel may be the safest form of avoiding action the presence of other vessels, the proximity of navigational hazards and other factors must obviously be taken into account in deciding how to manœuvre. If a hampered vessel takes action to avoid a give-way vessel approaching from fine on the starboard bow, which fails to keep out of the way, it may be safer to make a substantial turn to starboard. The give-way vessel is not required to avoid crossing ahead in this case and is likely to turn to starboard, especially by day when she may have failed to recognise the shapes displayed by the hampered vessel.

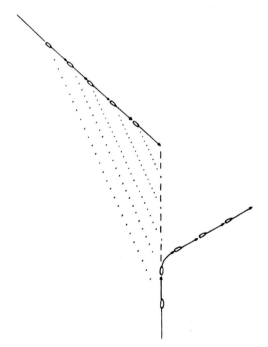

When vessels are in sight of one another a power-driven vessel which alters course to port or to starboard, or operates astern propulsion, is required to indicate the manœuvre by the whistle signals prescribed in Rule 34(a) and may supplement the sound signal with the light signal referred to in Rule 34(b). It is particularly important for both the give-way vessel and the stand-on vessel to make such signals, when taking action at a relatively late stage, in order to reduce the possibility of conflicting action being taken by the other vessel.

In the *Angelic Spirit–Y Mariner,* 1994, it was held that the vessels were crossing so as to involve risk of collision when they came into sight of one another on courses of 307° and 143° and that it was the duty of the *Y Mariner* to keep out of the way.

The *Angelic Spirit* was held partly to blame as she altered course only 20° to starboard when the vessels were about two miles apart. Mr Justice Clarke said:

It is not suggested that the time had come for action to be taken under Rule 17(b) when the ships were two miles apart. It is, however, said that action was permitted under Rule 17(a)(ii) because it had by then become apparent that Y Mariner *was not taking appropriate action as the give-way ship under Rule 15. In these circumstances I have asked the Elder Brethren whether that time had come on the facts found above and, if so, what action should have been taken. They have advised me that the time had come when it was permissible to take action, but that the action taken was not sufficient. They have further advised me that the action which should have been taken was a bold alteration of course accompanied by an equally bold reduction in speed and the appropriate sound signal to indicate action being taken. In this case, if action was taken, a 40° alteration of course to starboard, one short blast and action to stop engines was necessary in view of the highly dangerous close-quarters situation that had been allowed to develop. I accept that advice. It follows that the* Angelic Spirit *was at fault for failing to take proper action and for taking insufficient action instead.*

In the *Lok Vivek–Common Venture*, 1995, the *Lok Vivek* was held to be the stand-on vessel in a fine crossing situation. In considering whether the *Lok Vivek* was partly to blame and, if so, to what extent, Mr Justice Clarke said:

The Lok Vivek *went hard to starboard when the vessels were less than a mile apart. The question is whether she should have taken any other action and if so when. I have asked the Elder Brethren to assume the following facts. The vessels were approaching at similar speeds of 12.5 to 13 knots. They were crossing at an angle of 8° with the* Common Venture *bearing, say, 5° on the port bow of the* Lok Vivek. *On those assumptions I have asked the Elder Brethren what if any action the* Lok Vivek *should have taken as a matter of good seamanship in the light of Rule 17 of the regulations. They have advised me as follows.*

1. *When the* Common Venture *was distant about 2 to 3 miles, the* Lok Vivek *should have made a bold alteration of course to starboard as permitted by Rule 17(a)(ii) of the regulations.*
2. *When the* Common Venture *was distant about a mile, the* Lok Vivek *should have put her engines full astern and her wheel hard to starboard in accordance with Rule 17(b) of the regulations.*

I accept that advice. I observe in this connection that the time referred to in paragraph 1 above was about the time that the second officer was trying to contact the Common Venture *by VHF because he was worried that she was not taking appropriate action as the give-way ship in accordance with the regulations. In my judgment, as a matter of good seamanship in the light of Rule 17(a)(ii), he should have made a bold alteration of course to starboard. Finally, he should have put the engines full astern and the wheel hard to starboard when the ships were about a mile apart, whereas the only action he took was to put the wheel hard to starboard as set out above. In failing to take that action he was in breach of Rule 17(b).*

In the *Topaz–Irapua,* 2003, the *Topaz* was held to be the stand-on vessel in a crossing situation. In considering what action should have been taken by *Topaz* under Rule 17(a)(ii) Mr Justice Gross asked the assessors the following questions and received the following answers:

Q.1: On the assumptions that (1) the OOW on Topaz *had been observing* Irapua *from the time when the vessels were about 12 miles apart; (2)* Irapua *was and remained on a steady bearing; (3)* Irapua *had apparently taken no action to avoid a collision; and (4) the two vessels were closing rapidly; at what stage, if any, did good seamanship require* Topaz *(as the stand-on vessel) to take action to avoid collision by her man-œuvre alone under Rule 17(a)(ii) (before the stage at which Rule 17(b) applied)?*

A: By the time when the vessels were about three miles apart; i.e., ranges C-8 to C-10.

Q.2: What manoeuvre should have been undertaken by Topaz *at the time referred to in the answer to Question 1?*

A: A bold alteration of course to starboard of at least 30 deg.; such an alteration of course would have avoided the collision; an alteration of course of 10 deg. was inadequate. The Judge accepted that advice.

In the *Koscierzyna–Hanjin Singapore* the *Hanjin Singapore* was overtaking the *Koscierzyna,* approaching from directly astern, on the same course. The speed of the *Hanjin Singapore* was 21 knots, the speed of the *Koscierzyna* was 10.5 knots. Neither vessel took any action before the collision. It was held in the Court of Appeal (1995) that *Hanjin Singapore* was mainly to blame but that *Koscierzyna* should have altered course about 20°–30° (preferably to port) when the range had decreased to about one mile. *Koscierzyna* was held to be 15% to blame.

Compulsory action by the stand-on vessel

When the stand-on vessel finds herself so close that collision cannot be avoided by the give-way vessel alone she is required to take action. The distance between the two vessels at the moment when action becomes compulsory for the stand-on vessel will vary with the direction and speed of approach and will also depend on the give-way vessel's manœuvring characteristics. In a crossing situation this distance will usually be about four times the length of the give-way vessel.

As it is difficult to determine exactly how close the give-way vessel could approach before she is unable to avoid collision by her own action alone, the stand-on vessel should preferably take action before reaching this stage. An alteration of course to starboard to avoid a vessel approaching from the port bow could be a dangerous manœuvre if there is insufficient time to get clear. In the open sea it is suggested that a stand-on vessel should not allow a give-way vessel to approach to a distance of less than about twelve times her own length in a crossing situation without taking avoiding action.

When the vessels are so close that collision cannot be avoided by the give-way vessel alone the stand-on vessel is required to take such action as will best aid to avoid collision. Rule 17(c) does not apply at this stage, a power-driven vessel is permitted to turn to port for another power-driven vessel on the port bow. Turning towards the other vessel may be the best action to take at close quarters if one vessel appears likely to strike the other abaft the beam, as shown in the diagrams.

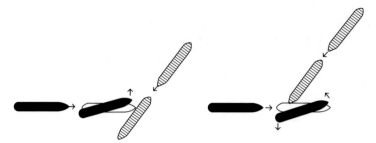

The above diagrams illustrate crossing cases in which the best helm action for the stand-on vessel to take to avert collision would be to turn to port.

When collision with another vessel is considered to be inevitable, the foremost concern of the officer must be to manœuvre his ship so as to reduce the effect of collision as much as possible. The consequences are likely to be most serious if one vessel strikes the other at a large angle near the mid length. The engines should be stopped, and the helm should be used so as to achieve a glancing blow rather than a direct impact. The damage would probably be the least serious if the impact is taken forward of the collision bulkhead. When a vessel is approaching on the port bow an alteration to starboard may well be the worst possible action to take.

Obligation of the give-way vessel

A disadvantage of permitting the stand-on vessel to take action to avoid collision by her manœuvre alone is that the give-way vessel may be tempted to wait in the hope that the stand-on vessel will keep out of the way. The purpose of Rule 17(d) is to emphasise that the give-way vessel is not relieved of her obligation to take early and substantial action to achieve a safe passing distance by the provisions of Rule 17(a)(ii). A stand-on vessel is not permitted to manœuvre until it becomes apparent that the give-way vessel is not taking appropriate action in compliance with the Rules. The give-way vessel should take positive action in ample time so that the stand-on vessel can maintain her course and speed. If the stand-on vessel takes action in accordance with Rule 17(a)(ii) the give-way vessel is not relieved of her obligation to keep out of the way and to achieve a safe passing distance.

The four stages in a collision situation

When two vessels in sight of each other are approaching with no change of compass bearing, so that when there is risk of collision one of them is required to keep out of the way by a Rule from Section II, there may be four stages relating to the permitted or required action for each vessel:

1. At long range, before risk of collision exists, both vessels are free to take any action.
2. When risk of collision first begins to apply the give-way vessel is required to take early and substantial action to achieve a safe passing distance and the other vessel must keep her course and speed.
3. When it becomes apparent that the give-way vessel is not taking appropriate action in compliance with the Rules the stand-on vessel is required to give the whistle

signal prescribed in Rule 34(d) and is permitted to take action to avoid collision by her manœuvre alone, but a power-driven vessel must not alter course to port to avoid another power-driven vessel crossing from her own port side. The give-way vessel is not relieved of her obligation to keep out of the way.

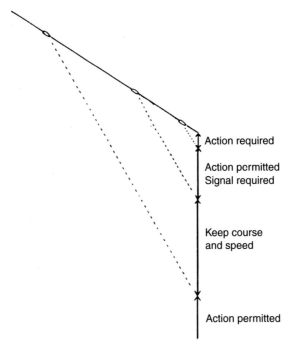

Action required

Action permitted
Signal required

Keep course
and speed

Action permitted

4. When collision cannot be avoided by the give-way vessel alone the stand-on vessel is required to take such action as will best aid to avoid collision.

The distances at which the various stages begin to apply will vary considerably. They will be much greater for high speed vessels involved in a fine crossing situation. For a crossing situation involving two power-driven vessels in the open sea it is suggested that the outer limit of the second stage might be of the order of 5 to 8 miles and that the outer limit for the third stage would be about 2 to 3 miles.

RULE 18

Responsibilities between vessels

Except where Rules 9, 10 and 13 otherwise require:

(a) A power-driven vessel underway shall keep out of the way of:
    (i)   **a vessel not under command;**
    (ii) **a vessel restricted in her ability to manœuvre;**
    (iii) **a vessel engaged in fishing;**
    (iv) **a sailing vessel.**

(b) **A sailing vessel underway shall keep out of the way of:**
   (i)   **a vessel not under command;**
   (ii) **a vessel restricted in her ability to manœuvre;**
   (iii) **a vessel engaged in fishing.**
(c) **A vessel engaged in fishing when underway shall, so far as possible, keep out of the way of:**
   (i)   **a vessel not under command;**
   (ii) **a vessel restricted in her ability to manœuvre.**
(d) (i) **Any vessel other than a vessel not under command or a vessel restricted in her ability to manœuvre shall, if the circumstances of the case admit, avoid impeding the safe passage of a vessel constrained by her draught, exhibiting the signals in Rule 28.**
   (ii) **A vessel constrained by her draught shall navigate with particular caution having full regard to her special condition.**
(e) **A seaplane on the water shall, in general, keep well clear of all vessels and avoid impeding their navigation. In circumstances, however, where risk of collision exists, she shall comply with the Rules of this Part.**
(f) (i) **A WIG craft when taking-off, landing and in flight near the surface shall keep well clear of all other vessels and avoid impeding their navigation;**
   (ii) **a WIG craft operating on the water surface shall comply with the Rules of this Part as a power-driven vessel.**

COMMENT:

Except where Rules 9, 10 and 13 otherwise require

Sailing vessels, vessels of less than 20m in length and vessels engaged in fishing, must comply with Rules 9(b) and 9(c), respectively, when in a narrow channel, and with Rules 10(j) and 10(i), respectively, when in a traffic lane. Any vessel about to cross a narrow channel or fairway must comply with Rule 9(d). The above Rules address the requirement not to impede the passage or safe passage of another vessel. A sailing vessel, a vessel of less than 20m in length or a vessel engaged in fishing, is required to take early action to allow sufficient sea room for the safe passage of a vessel navigating within a narrow channel, or following a traffic lane, and is not relieved of this obligation if risk of collision develops. A power-driven vessel, navigating within a narrow channel or following a traffic lane, is not entitled to keep her course and speed if approaching a sailing vessel, or vessel engaged in fishing, so as to involve risk of collision. As specified in Rule 8(f)(iii) such a vessel must comply with the Rules of this part, which includes Rule 18 (see page 47).

Any vessel which is overtaking any other vessel is required to keep out of the way of the vessel being overtaken. Rule 13 overrides Rule 18. A vessel which cannot easily alter her course should normally be able to reduce her speed. A vessel engaged in a special operation which cannot conveniently alter course or speed could request the other vessel to keep out of the way but must take avoiding action if the request is not complied with. The vessel being overtaken could comply with such a request as she is permitted by Rule 17(a)(ii) to take action to avoid collision by her own manœuvre when it becomes apparent that the overtaking vessel is not taking appropriate action.

Categories of vessels

A vessel may only be justified in regarding herself as falling within a certain category of privileged vessels for the purpose of Rule 18 if she satisfies the conditions of the relevant definition of Rule 3 and is also showing the lights or shapes prescribed in the appropriate Rule of Part C. A vessel engaged in a towing operation is not privileged with respect to other vessels unless she is severely restricted in her ability to deviate from her course.

Some vessels may not be easily identified as being of a special category and, by day, their shapes may not be sighted and recognized by other vessels in time for early avoiding action to be taken. This is more likely to apply if the hampered vessel is proceeding at high speed and the two vessels are meeting nearly end-on. In such circumstances the privileged vessel must make the signals prescribed in Rule 34(d) and take avoiding action, if possible, in accordance with Rule 17(a)(ii) (see pages 80–81).

Action to be taken when Rule 18 applies

A vessel required to keep out of the way by this Rule must take early and substantial action in accordance with Rule 16. She is not required to avoid crossing ahead but action must be such as to result in passing at a safe distance. It may be difficult for a hampered vessel to take effective avoiding action to avoid another vessel in a more privileged category but provision is made for such cases in the wording of Rule 18(c) and (d).

The privileged vessel is required to keep her course and speed in accordance with Rule 17(a)(i), so far as she is able to do so. The give-way vessel must take into account the possibility that the nature of the work being carried out, or the special circumstances which apply, may make it impossible for the stand-on vessel to keep both course and speed.

Power-driven vessel

A power-driven vessel is required to keep out of the way of all other types of vessel mentioned in Rule 18(a), except where Rules 9, 10 and 13 otherwise require, and must avoid impeding the safe passage of a vessel which is constrained by her draught and exhibiting the signals prescribed in Rule 28. These requirements apply when she is underway; a power-driven vessel lying stopped but 'under command' must comply with Rule 18.

Power versus sail

Although a power-driven vessel is required to keep out of the way of a sailing vessel when there is risk of collision, small yachts sailing for pleasure and making frequent alterations of course should keep well clear of large power-driven vessels so that it will not be necessary for the latter to take avoiding action. If a power-driven vessel does not take early action to keep out of the way it should usually be possible for a yacht to avoid collision by her manœuvre alone in accordance with Rule 17(a)(ii).

When the sidelight of a sailing vessel is sighted at night it may be useful to take the true direction of the wind into account so as to determine the approximate aspect of the vessel and to find out how she is sailing. Most sailing vessels can sail up to about 4 points (45°) of the wind. Some examples are given in the diagrams.

Responsibilities of a sailing vessel

A sailing vessel must keep out of the way of the vessels mentioned in Rule 18(b) and of a power-driven vessel which she is overtaking. A sailing vessel must also avoid impeding the safe passage of a vessel restricted to a narrow channel, following a traffic lane or constrained by her draught, in accordance with Rules 9(b), 10(j) and 18(d) respectively (see page 47).

Vessels engaged in fishing

A vessel engaged in fishing must, so far as possible, keep out of the way of a vessel not under command and a vessel restricted in her ability to manœuvre. She is also required to avoid impeding the safe passage of a vessel constrained by her draught. However, some vessels engaged in fishing may be unable to manœuvre as required by the Rules so that they are, in effect, 'not under command'.

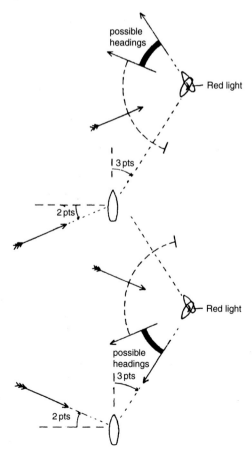

Power-driven vessels and sailing vessels must keep out of the way of vessels engaged in fishing, if risk of collision exists, and should also keep well clear of their nets or gear. A vessel fishing with nets extending more than 150 metres horizontally is required to show a white light or cone in the direction of the nets by Rule 26(c)(ii).

Drift net fishing vessels lay their nets in a continuous line extending for a considerable distance upwind. The nets may be set close to the surface and are supported by buoys at distances of approximately 40 metres apart. This type of gear is no longer in common use.

Vessels engaged in trawling may be dragging a net through the water at fairly high speeds, possibly in the region of 12 knots. Some trawlers drag a net along or near the bottom of the sea but others use the floating or mid-water trawl so other vessels should not approach closer than one mile astern.

Seine nets are commonly used in some areas. The fishing vessel first puts down a buoy then moves on a triangular path paying out rope, net, then more rope and returning to the buoy. The net is subsequently hauled in by winch, the whole operation taking two to three hours. The nets may be fairly close to the surface and can extend for over a mile from the fishing vessel so other vessels should keep well clear.

Line fishing vessels lay out long lines, with large numbers of hooks attached to them, along the sea bed. The lines are not likely to trouble other vessels which pass close by but the fishing vessel may be severely restricted in her ability to manoeuvre.

Vessels not under command

A vessel not under command could be either power-driven or under sail. The various circumstances in which a vessel may be considered not under command are discussed on pages 12–13. A vessel not under command may be making appreciable way through the water but having difficulty with steering so other vessels should keep well clear.

Vessels restricted in their ability to manoeuvre

Some of the vessels which are classed as being restricted in their ability to manoeuvre may be proceeding at relatively high speeds. This would apply especially to an aircraft carrier engaged in the launching or recovery of aircraft. The course and speed of such a vessel is governed by the force and direction of the wind. Ships engaged in replenishment at sea frequently proceed at speeds of 12 to 15 knots.

A vessel which is restricted in her ability to manoeuvre may not be justified in proceeding at high speeds in congested waters or when approaching yachts, vessels engaged in fishing and other low speed vessels. The manoeuvrability of the vessel with special reference to stopping distance and turning ability is one of the factors to be taken into account in determining what is a safe speed.

Vessels constrained by their draught

Any vessel, except a vessel not under command or restricted in her ability to manoeuvre, must avoid impeding the safe passage of a vessel constrained by her draught and exhibiting the signals mentioned in Rule 28. The words 'if the circumstances of the case admit' are included in Rule 18(d) to take account of the fact that some vessels, particularly vessels engaged in fishing, may be unable to take effective avoiding action in sufficient time.

Rule 8(f) must be taken into account by a vessel complying with Rule 18(d)(i). A vessel required to avoid impeding the safe passage of a vessel constrained by her draught must, if the circumstances of the case admit, take early action to allow sufficient sea room for the safe passage of the other vessel (see pages 43–45).

Early action can only be taken if the circumstances of the case admit. It may not be possible to recognise the lights or shape exhibited by a vessel constrained by her draught at sufficient range to enable action to be taken before risk of collision develops. However, as stated in Rule 8(f)(ii), the vessel which is required not to impede is not relieved of this obligation when there is risk of collision but when taking action must have full regard to the action which may be required by the Steering and Sailing Rules.

In a crossing situation in which a power-driven vessel has a vessel constrained by her draught on her own port side the power-driven vessel must, if the circumstances of the case admit, take early action to allow the safe passage of the other vessel. If the signals are not recognised at long range so that risk of collision develops the vessel constrained by her draught will become the give-way vessel, but the power-driven vessel should, if necessary, take action in accordance with Rules 8(f)(ii) and 17(a)(ii). She should avoid altering course to port in accordance with Rule 17(c).

Rule 18(d)(ii) requires a vessel constrained by her draught to navigate with particular caution having full regard to her special condition. According to the Rule 3(h) definition such a vessel is severely restricted in her ability to deviate from the course she is following. It is also probable that a vessel constrained by her draught will be severely limited in her ability to change her speed. Among the factors to be taken into account in determining a safe speed are the manoeuvrability of the vessel with special reference to stopping distance and turning ability in the prevailing conditions and the draught in relation to the available depth of water. It is, therefore, doubtful whether a vessel constrained by her draught would be justified in proceeding at full speed when other vessels are in the vicinity.

The following Guidance on the application of Rule 18(d) has been approved by IMO:

'Clarification of the relation between Rule 18(d) and the Rules of Part B, Sections II and III. *A vessel constrained by her draught shall, when risk of collision with another vessel in a crossing or head-on situation exists, apply the relevant Steering and Sailing Rules as a power-driven vessel. She should, when showing the signals prescribed by Rule 28, have her engines ready for immediate manoeuvre and proceed at safe speed as required by Rule 6.*'

Local rules

Rule 1(b) permits special rules to be made for specific areas, by an appropriate authority, which take precedence over the International Regulations for Preventing

Collisions at Sea. Mariners should be aware that some authorities have made special rules which give additional privilege to a vessel constrained by her draught so that she does not become a give-way vessel and that such rules may be applicable in coastal waters which are within the territorial limits of the nation concerned. The Sailing Directions and other publications should be consulted for details of such special rules (see also page 8).

Situations involving two hampered vessels

In the case of two hampered vessels approaching one another in meeting or crossing situations so as to involve risk of collision in which the degree of responsibility is not established, each vessel should take whatever action she can to avoid collision. This would apply in the case of a vessel restricted in her ability to manœuvre meeting a vessel which is not under command, or when a hampered vessel approaches another vessel of the same category. Alterations of course should preferably be to starboard, in accordance with the principles of Rules 14, 15 and 17(c).

Seaplanes, hovercraft and hydrofoils

Rule 18(e) refers to seaplanes which must, in general, keep well clear of all vessels and avoid impeding their navigation, but must comply with the Rules when risk of collision exists. For the purpose of the Rules hovercraft and hydrofoils are not classed as seaplanes, even when operating in the non-displacement mode, but are to be considered as power-driven vessels. It was decided that they should not be required to keep out of the way of all other vessels as air-cushion vessels are not always capable of achieving very high speeds. However, it might be considered to be an act of good seamanship, in compliance with Rule 2(a), for hovercraft and hydrofoils proceeding at high speed to take early action to keep well clear of all shipping.

Air-cushion vessels operating in the non-displacement mode are very susceptible to wind effects. Such vessels may have a drift angle of as much as 45°, so their navigation lights may give a false indication of the direction of travel. It is mainly for this reason that all air-cushion vessels are required by Rule 23(b) to exhibit an all-round flashing yellow light in addition to the lights prescribed for power-driven vessels underway.

WIG craft

Paragraph (f) was added to Rule 18 by the 22nd Assembly of IMO in 2001. This paragraph refers to WIG craft, which are required when taking-off, landing and in flight near the surface to keep well clear of all other vessels and avoid impeding their navigation. This requirement also applies to a WIG craft taking-off, landing or in flight near the surface when there is risk of collision with another vessel. When operating on the water surface, not taking-off or landing, a WIG craft must comply with the Rules as a power-driven vessel.

Section III – Conduct of Vessels in Restricted Visibility

RULE 19

Conduct of vessels in restricted visibility

(a) This Rule applies to vessels not in sight of one another when navigating in or near an area of restricted visibility.

(b) Every vessel shall proceed at a safe speed adapted to the prevailing circumstances and conditions of restricted visibility. A power-driven vessel shall have her engines ready for immediate manœuvre.

(c) Every vessel shall have due regard to the prevailing circumstances and conditions of restricted visibility when complying with the Rules of Section I of this Part.

(d) A vessel which detects by radar alone the presence of another vessel shall determine if a close-quarters situation is developing and/or risk of collision exists. If so, she shall take avoiding action in ample time, provided that when such action consists of an alteration of course, so far as possible the following shall be avoided:

　(i)　an alteration of course to port for a vessel forward of the beam, other than for a vessel being overtaken;

　(ii)　an alteration of course towards a vessel abeam or abaft the beam.

(e) Except where it has been determined that a risk of collision does not exist, every vessel which hears apparently forward of her beam the fog signal of another vessel, or which cannot avoid a close-quarters situation with another vessel forward of her beam, shall reduce her speed to the minimum at which she can be kept on her course. She shall if necessary take all her way off and in any event navigate with extreme caution until danger of collision is over.

COMMENT:

In or near an area of restricted visibility

The term 'restricted visibility' is defined in Rule 3(1), (see page 6). Rule 19 applies not only when a vessel is navigating in an area of restricted visibility but also when she is near such an area. A vessel which is approaching an area of restricted visibility, or which has such an area on one side, must comply with Rule 19 and must also give the sound signals prescribed in Rule 35.

Gladiator–St Paul

Over and over again we have had cases in this Court where a vessel not herself in a fog has been blamed because, seeing a fog ahead, she has not taken precautions, so that her speed shall be off when she enters the fog. There is a difference in snow, but the same kind of considerations apply. If there is a thick snowstorm ahead, so that nothing can be seen in it, good seamanship requires there should be a moderate rate of speed, so as to approach that place under proper control. (Sir Gorell Barnes, 1909)

Not in sight of one another

Rule 19 in Section III applies to vessels not in sight of one another in restricted visibility whereas the Rules of Section II apply to vessels in sight of one another whether or not the visibility is restricted. As soon as vessels navigating in or near an area of restricted visibility come in sight of one another they must comply with the Rules of Section II. Vessels not in sight of one another should not give the manœuvring and warning signals prescribed in Rule 34.

Safe speed

The 1960 Regulations required every vessel to go at a moderate speed in restricted visibility. The term 'safe speed' has now been substituted as Rule 6 applies to every vessel at all times, but previous Court interpretations of the term 'moderate speed' are still relevant when considering what is meant by a safe speed in restricted visibility (see pages 18–20).

The extent of visibility at which it first becomes necessary to reduce speed will depend upon the speed of the ship, her stopping power, the traffic in the vicinity and other factors. If the visibility is less than 5 miles it would be prudent for any vessel to, at least, have the engines on stand-by as fog can develop rapidly.

A reduction of speed is not necessarily required due to a sudden onset of a heavy rainstorm. If the visibility was good before the rain started and the rain is not expected to last long a vessel may be justified in maintaining speed in the light of the prevailing circumstances. Radar can be used to indicate the extent and movement of a rainstorm and to detect large vessels within and beyond the rain area but small craft may not be detected in heavy rain so the speed should be reduced if the rainfall is likely to continue for more than a few minutes.

The main factors to be taken into account in determining safe speed are listed in Rule 6. When the visibility is restricted the other most important factors will usually be traffic density, own ship's manœuvrability and the efficiency of the radar equipment. In the open sea, with little or no traffic in the vicinity, a relatively high speed may be appropriate for the prevailing circumstances and conditions provided a proper radar watch is being kept and the engines are ready for immediate manœuvre, but even a vessel with good stopping power using a sophisticated collision avoidance system would not be justified in proceeding at high speed in dense fog through congested waters or areas where small craft and ice are likely to be encountered.

Some masters may be reluctant to make appreciable reductions of speed in restricted visibility because of pressure to maintain schedules. The attitude of owners and marine superintendents is likely to have been affected by decisions of the Courts in *The Lady Gwendolen* case.

On the 10th November, 1961, a collision occurred in dense fog between the *Freshfield* and *The Lady Gwendolen,* when the *Freshfield* was lying at anchor in the River Mersey. At the Formal Investigation held in March 1962, it was found that the collision was solely caused by the wrongful act or default of the master of *The Lady Gwendolen*, and his certificate was suspended.

In an action brought before the Admiralty Court in June 1964, the owners of *The Lady Gwendolen* sought to limit their liability. It was held that the owners were guilty of actual fault and were unable to limit. This judgment was upheld by the Court of Appeal.

In the Admiralty Court Mr Justice Hewson said:

After weighing up this case and the evidence and the circumstances with what I hope is all the care of which I am capable, I am driven to the conclusion that a total lack of a sense of the urgency of the problem posed by radar navigation in fog in Captain Meredith was a contributory cause of the collision, and this sense of urgency and importance should have been instilled in him from the highest level.

In the Court of Appeal Lord Justice Sellers said:

A primary concern of a shipowner must be safety of life at sea. That involves a seaworthy ship, properly manned, but it also requires safe navigation. Excessive speed in fog is a grave breach of duty, and shipowners should use all their influence to prevent it. In so far as high speed is encouraged by radar the installation of radar requires particular vigilance of owners.

Lord Justice Willmer said:

In the course of his evidence Captain Meredith was cross-examined at some length on his log records of various previous voyages undertaken in conditions of fog. This led in the end to an admission by Captain Meredith that he had for years habitually navigated his vessel in fog at excessive speed. Mr Robbie (the marine superintendent) gave evidence to the effect that on a number of occasions he had spoken to Captain Meredith, and to the masters of the other vessels, about the problem of navigation in fog with the aid of radar. This evidence of Mr Robbie was, however, denied by Captain Meredith, and was disbelieved by the learned judge. It became quite apparent from the cross-examination of Mr Robbie that, although all the ships' logs were regularly submitted to him, he had signally failed to check the records contained therein with a view to ascertaining how The Lady Gwendolen *was being navigated in fog. It would not have required any very detailed examination of the engine room records in order to ascertain that* The Lady Gwendolen *was frequently proceeding at full speed at times when the deck log was recording dense fog. Yet this fact appeared never to have been detected by Mr Robbie, and consequently was never brought to the attention of Captain Meredith.*

It was said that the lack of managerial control shown in this case was to be contrasted with the practice prevailing in other companies where 'the management had evolved an effective system for keeping a check on the way in which the companies' vessels were navigated'. However, it was not suggested that any pressure was exerted upon the master to keep his schedule. It was stated that the radar problem was one of such serious import as to merit and require the personal attention of the owners, but that in this case no steps had been taken to ensure that the masters used their radar in a proper manner.

Ready for immediate manœuvre

A power-driven vessel is required to have her engines ready for immediate manœuvre in restricted visibility; this will apply even in the open ocean. For many vessels placing the engines on stand-by involves some reduction of speed and loss of economy in fuel but this must be accepted in the interests of safety. As it may take several minutes to prepare the engines for immediate manœuvre the engineers should be given as much notice as possible when it seems likely that the visibility will become restricted.

Complying with the rules of Section I

Rule 19(c) emphasises the need to take the circumstances of restricted visibility into account when complying with the Rules of Section I of Part B. In addition to Rule 6 which relates to safe speed this will apply particularly to Rules 5, 7 and 8 dealing with look-out, risk of collision and avoiding action. The Rules relating to navigation in narrow channels and traffic separation schemes also apply in all conditions of visibility.

In order to keep a good look-out in restricted visibility it will be necessary to have a man posted on look-out duty by day as well as by night and the radar should be kept under practically continuous observation by a competent person. The use of radar will be essential, if fitted and operational, to determine whether risk of collision exists with a vessel detected but not in sight in restricted visibility. More substantial alterations of course will be necessary to avoid collision with a vessel which is not in sight so that the manœuvres will be readily apparent on the other vessel's radar screen, as required by Rule 8(b). The effectiveness of avoiding action must be carefully checked by radar observation if the other vessel is not in visual sight.

Detection by radar alone

Rule 19(d) applies to a vessel which detects another vessel, in restricted visibility, by radar alone, i.e., without sighting her visually or hearing her fog signal. The Rules of Section II apply to vessels in sight of one another and Rule 19(e) applies when a fog signal is heard and there is possible risk of collision. If the vessel detected comes into visual sight, or if a fog signal is heard forward of the beam, the appropriate Rule must be complied with. It is essential to keep a good look-out by sight and hearing in addition to making proper use of the radar.

Determine risk of collision

A vessel which detects another vessel by radar alone in restricted visibility is required to determine whether a close quarters situation is developing and/or risk of collision exists. Rule 7(b) also requires that proper use be made of radar equipment to obtain early warning of risk of collision, and that radar plotting or equivalent systematic observation should be carried out. Assumptions must not be based on scanty information (see pages 32–33).

A close quarters situation

Rules 8(c), 19(d) and 19(e) refer to a close quarters situation. The distance at which a close quarters situation first applies has not been defined in miles, and is not likely to be, as it will depend upon a number of factors. The 1972 Conference considered the possibility of specifying the distance at which it would begin to apply but after a lengthy discussion it was decided that this distance could not be quantified.

Grepa–Verena

It leaves open to argument what is meant by the phrase 'close quarters situation'. That, I think, must depend upon the size, characteristics and speed of the ships concerned. I think, however, that in the case of ships of the class that we have here it must mean a quite substantial distance, and, I would venture to think, a distance measurable in miles rather than in yards. (Lord Justice Willmer, 1961)

In restricted visibility, in the open sea, a close quarters situation is generally considered to begin to apply at a distance of at least 2 miles in any direction forward of the beam as this is the typical range of audibility for the whistle of a large vessel in still conditions (see Annex III(1)(c)). A minimum distance of 3 miles is sometimes suggested when determining whether a close quarters situation is developing as allowance should be made for the effects of errors in radar observations, especially at long range. However, distances of less than 2 miles may be considered sufficient when proceeding at reduced speed in congested waters, when in an overtaking situation, or, when a vessel is expected to pass astern.

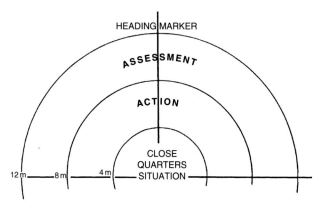

Taking avoiding action in ample time

Avoiding action must be taken if a close quarters situation is developing and/or risk of collision exists. It is not necessary to take avoiding action if a vessel is expected to pass at a close distance but there is no risk of collision as, for instance, when two vessels are proceeding in opposite directions on their correct sides within a narrow channel.

Rule 8(a) requires avoiding action to be taken in ample time in all conditions of visibility. When the visibility is restricted it is generally necessary to take action to avoid a close quarters situation at an earlier stage. However, action should not be taken without first making a full assessment of the situation. Rule 7(c) states that assumptions shall not be made on the basis of scanty information, especially scanty radar information.

As a general guide it has been suggested that, using a 12 mile range scale in the open sea, radar observations should be assessed as an approaching target crosses the outer one third of the screen to see whether a close quarters situation is developing. If so substantial action should be taken before the target reaches the inner one third of the screen.

Rule 19(d) requires avoiding action to be taken in ample time if a close quarters situation is developing with a vessel approaching from any direction. A vessel which is being overtaken is not required, or even permitted, to keep her course and speed when a close quarters situation is developing. The Rules of Section II only apply to vessels in sight of one another. However, when a vessel is approaching from abaft the beam the relatively low rate of approach means that action can be taken at shorter range and yet be made in ample time.

When action consists of an alteration of course

It was recommended in the Annex to the 1960 Rules that in order to avoid a close quarters situation in restricted visibility an alteration to starboard is generally preferable to an alteration to port, particularly for vessels approaching apparently on opposite or nearly opposite courses. This recommendation has subsequently been considered to have been insufficient for the purpose of discouraging vessels from turning to port in meeting or crossing situations so it was made mandatory, by the 1972 Conference, to avoid altering course to port for a vessel forward of the beam, except when overtaking. Rules 14, 15 and 17(c) virtually impose a similar restriction on power-driven vessels in sight of one another which are meeting or crossing so as to involve risk of collision.

An alteration of course to port to avoid a vessel being overtaken is permitted as an alternative to an alteration to starboard, or change of speed, whether the vessels are in sight or not. In the open sea a vessel which is overtaking should preferably take action to avoid a close quarters situation when the two vessels are several miles apart so that the vessel being overtaken will be relieved of her obligation to take avoiding action and will be less likely to make a conflicting manœuvre.

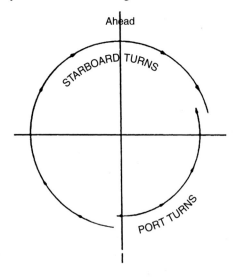

Alterations of course towards a vessel approaching from abeam or abaft the beam are to be avoided. This means that alterations of course to port should not be made to avoid a close quarters situation with a vessel approaching from any direction on the port side or from the starboard bow and that an alteration to starboard should not be made for a vessel approaching from the starboard beam or starboard quarter. An alteration in either direction is permitted when a vessel approaches from astern.

The purpose of the requirement to avoid turning to port when a close quarters situation is developing with a vessel forward of the beam is to reduce the possibility of conflicting action being taken by vessels on opposite or nearly opposite courses. A substantial alteration of course to port to avoid a vessel approaching from just forward of the starboard beam is a relatively safe manœuvre as the difference between courses is likely to be less than 90°. The line of demarcation between 'abeam' and 'forward of the beam' is not specified so an alteration to port to avoid a vessel approaching from within about two points of the starboard beam would not be a clear contravention of Rule 19(d). When vessels are in sight of one another a power-driven vessel is permitted to make a substantial alteration of course to port to avoid another power-driven vessel approaching from just forward of the starboard beam as such action would not involve crossing ahead of the other ship.

Provision is made in Rule 19(d) for exceptions to the restrictions on course changes by the inclusion of the words 'so far as possible'. However, if an alteration to port is decided upon, due perhaps to lack of sea room to starboard or to the presence of other vessels, it is especially important that it should be made as early as possible and that it should be a bold alteration when avoiding a close quarters situation with a vessel approaching from ahead or fine on the bow.

Change of speed to avoid a close quarters situation

Avoiding action must be taken if a close quarters situation is developing and there is risk of collision but a change of speed can be made as an alternative to, or in association with, an alteration of course.

A change of speed is generally more effective as a means of avoiding a close quarters situation with a vessel approaching from near the beam. In a meeting situation a reduction of speed is unlikely to have an appreciable effect on the distance of closest approach but it could be considered to be 'avoiding action' as it reduces the closing speed and gives more time for assessment and further action by both vessels (Rule 8(e)).

19(e) Where risk of collision does not exist

Determination of risk of collision is required by Rule 19(d) and by Rule 7. A series of radar ranges and bearings, together with a plot or equivalent systematic observation, indicating that it is safe to proceed will usually be necessary to justify continuing at a speed greater than bare steerage-way. The possibility that a fog signal may be heard from a different vessel to the one whose echo has been observed must also be taken into account. The direction and distance of sound signals can be misleading in fog. In the *Oakmore–Aras,* 1907, Sir Gorell Barnes put the following question to the Elder Brethren:

Were the indications such as to show her master, distinctly and unequivocally, that if both vessels continued to do what they appeared to be doing, they would pass clear without risk of collision?

If it has been determined that risk of collision does not exist a vessel is not required to reduce her speed to the minimum at which she can be kept on her course when a close quarters situation is developing, or on hearing a fog signal forward of her beam. This may apply, for instance, when crossing astern of a vessel being overtaken or when vessels proceeding in opposite directions on their correct sides of a narrow channel pass close enough to hear each other's fog signals. A ship's whistle can sometimes be heard at long distances.

Fog signal apparently forward of the beam

Although the Rule refers only to fog signals heard apparently forward of the beam it may be prudent to reduce speed if a signal appears to come from near or slightly abaft the beam. The direction of sound signals cannot be relied upon. It was held in one case (*Bremen–British Grenadier,* 1931) that a vessel should have stopped her engines, on the grounds of good seamanship, when several signals were heard just abaft the beam on the same bearing.

Fog signal reported

If the master or officer in charge is informed that a fog signal has been heard apparently forward of the beam when the vessel is proceeding at a speed appreciably above bare steerage way, the engines should be stopped or speed reduced, unless it has been determined that there is no risk of collision. A look-out posted forward may be able to hear a fog signal before it can be heard from the bridge.

Chusan–Protector

I see no excuse for the failure of the Master and pilot to act upon the report made to them by the third officer, when he informed them that he had heard the whistle of a vessel ahead. It seems to me that it is no excuse on the part of either pilot or master to say he did not hear it himself. If the officer of the watch, or the look-out, or anybody else, reports the hearing of a whistle from a vessel forward of the beam, it seems to me the imperative duty ... comes into force at once. (Mr Justice Willmer, 1955)

Signal of anchored vessel

Rule 19(e) applies when the fog signal of any other vessel is heard apparently forward of the beam. If the fog signal of a vessel at anchor is heard, apparently ahead, and the vessel has not previously been detected by radar, the engines should be stopped and the way taken off if necessary. This will also apply to signals heard from such vessels as wreck marking vessels.

Cannot avoid a close quarters situation

The speed must be reduced to the minimum at which a vessel can be kept on her course if a close quarters situation cannot be avoided, with a vessel forward of her beam. If a vessel is unable to avoid a close quarters situation, due perhaps to lack of sea room or to action taken by the other vessel, she must reduce speed in ample time without waiting for a close quarters situation to develop. The greater the initial speed the greater the range at which the speed should be reduced.

Shall reduce her speed

Any alteration of speed to avoid collision with a vessel not in sight should be large enough to be readily apparent to another vessel observing by radar (Rule 8(b)). Stopping the engines may be the most effective way of bringing the speed down. It may also provide greater opportunity for hearing the fog signals of the other vessel; this will have particular application to a vessel without operational radar.

A vessel may be justified in maintaining a speed greater than bare steerage way if the radar indicates that a close quarters situation is developing with a vessel approaching from just forward of the beam, or very broad on the bow which is expected to pass astern. The safest action may be to turn away from the other ship. Rule 2(b) permits departures from the Rules to be made in special circumstances.

Navigate with extreme caution

The term 'navigate with caution' was used in the 1960 and previous Regulations. For a vessel without operational radar which hears a fog signal forward at the beam it has generally been interpreted to mean that the way should at least be run off.

Union–Vulcano

She said that when she heard that whistle she went dead slow, and dead slow on that ship is accomplished by stopping her engines for a minute and then going on ahead again, then stopping and then going dead slow ahead, and it is said that that system of stopping and going dead slow is a compliance with Rule ... of stopping engines and navigating with caution in fog. I do not think that will do. I do not think it was intended to be so. I think that she intended to go on dead slow. Even if that was the true case, I do not think that stopping and going on again slow is a compliance with the Rule and to stop and navigate with caution. I think a compliance with the Rule is to stop your engines and get all the way off your ship for certain, and then go on again if you have heard a whistle from the other ship; and if you have heard nothing at all I doubt if you are justified in going on until you do. (Mr Justice Bateson, 1928)

Alterations of course should, in general, be avoided after hearing a fog signal forward of the beam, unless both the position and movement of the other vessel have been reasonably determined. There have been many Court decisions to this effect.

Miguel de Larrinaga–Hjelmaren

It is because it is so easy to be deceived by sounds heard in fog that it has been said in this Court time and time again that it is wrong for vessels, particularly vessels carrying a lot of headway, to alter course in fog merely on the faith of a fog signal. I only desire to add that I express that view, not only with all the force at my command, but with the additional authority of the Elder Brethren, who have advised me in this case in the same sense as their predecessors in many cases for generations have advised previous judges in this court. (Mr Justice Willmer, 1956)

The direction of sound signals may be misleading in fog.

Oakmore–Aras

... it is so well known – so absolutely well known – that it is impossible to rely upon the direction of whistles in a fog, that I do not think any man is justified in relying with certainty upon what he heard when the whistle is fine on the bows, like this was undoubtedly, and is not justified in thinking it is broadening unless he can make sure of it. That is the view I entertain very strongly, because, if it is well established that the direction of sound in a fog is a matter of uncertainty, it is no use trying to make it a certainty by saying you looked at the compass. (Sir Gorell Barnes, 1906)

A vessel which cannot avoid a close quarters situation with another vessel detected by radar forward of the beam should also avoid making a blind alteration of course when the other vessel is at short range and her course has not been ascertained. In each of the following cases both vessels altered course on the basis of insufficient radar information when in a close quarters situation.

Thorshovdi–Anna Salen

I find that both vessels, although at different times and in different circumstances, violated one of the cardinal rules of seamanship by altering course blindly, without having any precise knowledge of what the other vessel was doing. (Mr Justice Willmer, 1954)

Linde–Aristos

As regards alterations of course, I have found that both ships altered about the same time and about the same amount. It was argued for the defendants that the Linde was in better case because she altered to starboard rather than to port. I cannot see this. I have been advised by the Elder Brethren that any alteration of course at the time made, namely, before sighting, and without the course of the other ship having been properly ascertained, was unseamanlike. I accept that advice. I cannot see that there is any significant difference between the two ships in this respect. (Mr Justice Brandon, 1969)

Alterations of course are not always condemned by the Courts. An alteration may be justified if a sufficient number of fog signals, or radar observations, have given a reasonable indication of the position and movement of the other vessel.

Vindomora–Haswell

At the same time it appears also to me to be a principle of common sense and good seamanship that when two vessels are near together in a fog, and the one receives a sufficient indication of the position of the other, there is no rule, and there could be no rule, that the vessel which receives such an indication, and thereby has good reason for changing her course should not do so. (Lord Morris, 1890)

Sedgepool–Parthia

In the particular conditions and the particular locality where this collision happened, it would be impossible to say that either vessel was wrong for altering course to starboard, even though the other vessel was not in sight. This was a collision which, upon my findings, occurred in a narrow channel, and in those circumstances, I should be very slow to blame a ship which on hearing a fog signal from another vessel, apparently approaching in the opposite direction in the same channel, altered her course to starboard in an attempt to get more over to her proper side. (Mr Justice Willmer, 1956)

If an alteration of course is made for another vessel which has not been sighted visually, the signals prescribed in Rule 34 must not be used.

If necessary take all way off

The courts have held that vessels navigating without radar should have reversed their engines after hearing a fog signal forward of the beam in the following instances:

a. when the signal was heard for the first time in close proximity;
b. where the signal was heard dead ahead;
c. where the signals were narrowing on the bow;
d. where a vessel was seen to loom out of the fog but her course was not immediately apparent;
e. where a sailing vessel's fog signal was heard forward of the beam;
f. where the signal was that of a vessel at anchor and the tide was setting towards her.

However, the engines should not be put astern unnecessarily, especially full astern, if the engine noise may make it difficult to hear signals.

Monarch–Jaunty

I have always understood that one of the reasons why the Regulations require the stopping of the engines in fog, when a signal is heard from another ship, is so as to enable further signals to be heard the better. It appears to me that when there is any question of listening for signals one is creating the worst possible conditions for hearing them by working the engines at full speed astern. Moreover, the fact of taking drastic action like that cannot do other than cause a certain degree of diversion of attention. (Mr Justice Willmer, 1953)

A vessel navigating with radar which cannot avoid a close quarters situation with another vessel forward of her beam may also be expected to put the engines astern and take all her way off, especially when the other vessel is approaching from ahead or within about 30° of the bow. In taking such action, however, account must be taken of the effect of transverse thrust and/or wind action which may slew the vessel across the path of the oncoming ship. If this should occur as the vessel is coming to rest a short burst of ahead power with the rudder hard over may serve to keep the bow pointing towards the approaching ship.

It is a sound principle of collision avoidance to stop as rapidly as possible and face the danger when there is doubt as to which side any vessel approaching directly at a relatively high speed may attempt to pass by. Risk of collision is reduced as a vessel end-on presents a smaller target. Should there be a collision the effect is likely to be much less serious if the impact is taken forward of the collision bulkhead than if struck at a broad angle near the mid length.

Part C – Lights and shapes

RULE 20

Application

(a) Rules in this Part shall be complied with in all weathers.

(b) The Rules concerning lights shall be complied with from sunset to sunrise, and during such times no other lights shall be exhibited, except such lights as cannot be mistaken for the lights specified in these Rules or do not impair their visibility or distinctive character, or interfere with the keeping of a proper look-out.

(c) The lights prescribed by these Rules shall, if carried, also be exhibited from sunrise to sunset in restricted visibility and may be exhibited in all other circumstances when it is deemed necessary.

(d) The Rules concerning shapes shall be complied with by day.

(e) The lights and shapes specified in these Rules shall comply with the provisions of Annex I to these Regulations.

COMMENT:

All weathers

Even small vessels are expected to comply with the Rules concerning lights and shapes in all weather conditions. If any lights are lost or extinguished they must be replaced or repaired as soon as possible. Oil lanterns, which may be provided for use in emergencies, should be properly maintained and kept ready for use. A delay in attending to lights or shapes in severe weather conditions because of the danger to personnel may be justified but should be recorded in the official log book.

No other lights

A vessel of 100 metres or more in length is required to use the available working or equivalent lights to illuminate her decks when at anchor, by Rule 30(c), and a smaller vessel at anchor may use such lights. When weighing anchor the deck lights must be switched off, with the anchor lights, as soon as the anchor is out of the ground.

In the following extract from the judgment in the case of *Tojo Maru–Fina Italia* reference is made to Rule 1(b) of the 1954 Regulations which was almost identical to Rule 20(b) of the 1972 Regulations:

In my view the lights which the Fina Italia was exhibiting offended against that Rule in more than one respect. I received the impression from the evidence given … that it was not altogether unknown for tankers underway off Kuwait to be manœuvring with their deck lights burning. If there is any such habit in the port of Kuwait, I can

A Guide to the Collision Avoidance Rules. DOI: 10.1016/B978-0-08-097170-4.00003-9

only say that the sooner it is discontinued the better, because it is bound to increase the difficulty of navigation for other vessels. (Lord Justice Willmer, 1968)

Restricted visibility

The prescribed lights, if carried, must also be exhibited in restricted visibility from sunrise to sunset. The words 'if carried' are included as some vessels, such as ferries, are not fitted with navigation lights as their operations are restricted to daylight hours.

By day

Shapes must be exhibited by day, not merely from sunrise to sunset. A vessel required to show a day signal should exhibit both lights and shape(s) during the period of twilight.

RULE 21

Definitions

(a) 'Masthead light' means a white light placed over the fore and aft centreline of the vessel showing an unbroken light over an arc of the horizon of 225 degrees and so fixed as to show the light from right ahead to 22.5 degrees abaft the beam on either side of the vessel.

(b) 'Sidelights' means a green light on the starboard side and a red light on the port side each showing an unbroken light over an arc of the horizon of 112.5 degrees and so fixed as to show the light from right ahead to 22.5 degrees abaft the beam on its respective side. In a vessel of less than 20 metres in length the sidelights may be combined in one lantern carried on the fore and aft centreline of the vessel.

(c) 'Sternlight' means a white light placed as nearly as practicable at the stern showing an unbroken light over an arc of the horizon of 135 degrees and so fixed as to show the light 67.5 degrees from right aft on each side of the vessel.

(d) 'Towing light' means a yellow light having the same characteristics as the 'sternlight' defined in paragraph (c) of this Rule.

(e) 'All-round light' means a light showing an unbroken light over an arc of the horizon of 360 degrees.

(f) 'Flashing light' means a light flashing at regular intervals at a frequency of 120 flashes or more per minute.

COMMENT:

The main specifications of each type of light with regard to general position on the vessel, colour and arc of visibility are given in these definitions so as to avoid repetition in subsequent Rules. More detailed requirements concerning the position and characteristics of lights and shapes are given in Annex I.

Masthead light

Although the term 'masthead light' is used the text does not specifically require the light to be placed on a mast. Annex 1.2(f) states that masthead lights shall be so placed as to be above and clear of all other lights and obstructions.

Sidelights

Sidelights must be fitted with inboard screens in accordance with Annex I(5) to give a practical cut-off between 1° and 3° outside the prescribed sector, as specified in Annex I(9)(a). This means that the rays of light from the outer part of the filament, or wick, may cross the fore and aft line and be visible to a vessel approaching from an angle of up to 3° on the opposite bow. If some of the light were not permitted to show across the bow there would be a theoretical 'dark lane' ahead which could result in vessels meeting exactly end-on being unable to see each other's sidelights.

Sternlight

The sternlight is required to be placed 'as nearly as practicable at the stern' as some vessels such as towing vessels, stern trawlers and LASH vessels with an open stern could find it difficult or even impossible to carry a light at the stern.

Towing light

This light is prescribed only for a vessel engaged in towing another vessel from the stern.

The 1960 Regulations prescribed only three colours of light: red, green and white. In the 1972 Regulations both the towing light and the flashing light of an air-cushion vessel are required to be yellow and Annex II permits vessels fishing with purse seine gear to exhibit two yellow lights which flash alternately. Consideration was given to the possibility of using yellow for a stern light but tests indicated that yellow and white lights are not distinguishable unless they are in juxtaposition. The yellow towing light is required to be carried above the white stern light.

All-round light

It may not be possible to show an unbroken light over an arc of the horizon of 360°. This definition is further qualified by paragraph 9(b) of Annex I which requires all-round lights, other than anchor lights, to be so placed as not to be obscured within angular sectors of more than 6°.

Flashing light

The flashing light is prescribed only for air cushion vessels (Rule 23(b)) and WIG craft (Rule 23(c)). The rate of flashing is higher than that previously specified for hovercraft in the *Mariner's Handbook* and Notices to Mariners. The rate was made deliberately high to enable mariners to distinguish this light from the flashing lights of buoys and other aids to navigation which usually do not flash more than 60 times per minute. Submarines may also exhibit a flashing light.

RULE 22

Visibility of lights

The lights prescribed in these Rules shall have an intensity as specified in Section 8 of Annex I to these Regulations so as to be visible at the following minimum ranges:

(a) **In vessels of 50 metres or more in length:**
- a masthead light, 6 miles;
- a sidelight, 3 miles;
- a sternlight, 3 miles;
- a towing light, 3 miles;
- a white, red, green or yellow all-round light, 3 miles.

(b) **In vessels of 12 metres or more in length but less than 50 metres in length:**
- a masthead light, 5 miles; except that where the length of the vessel is less than 20 metres, 3 miles;
- a sidelight, 2 miles;
- a sternlight, 2 miles;
- a towing light, 2 miles;
- a white, red, green or yellow all-round light, 2 miles.

(c) **In vessels of less than 12 metres in length:**
- a masthead light, 2 miles;
- a sidelight, 1 mile;
- a sternlight, 2 miles;
- a towing light, 2 miles;
- a white, red, green or yellow all-round light, 2 miles.

(d) **In inconspicuous, partly submerged vessels or objects being towed:**
- a white all-round light, 3 miles.

COMMENT:

Visible

Paragraph 8 of Annex I gives a formula for calculating the luminous intensity of lights which takes account of visibility and atmospheric transmissivity.

RULE 23

Power-driven vessels underway

(a) **A power-driven vessel underway shall exhibit:**
   (i)   **a masthead light forward;**

(ii) a second masthead light abaft of and higher than the forward one; except that a vessel of less than 50 metres in length shall not be obliged to exhibit such light but may do so;

(iii) sidelights;

(iv) a sternlight.

(b) An air-cushion vessel when operating in the non-displacement mode shall, in addition to the lights prescribed in paragraph (a) of this Rule, exhibit an all-round flashing yellow light.

(c) A WIG craft only when taking-off, landing and in flight near the surface shall, in addition to the lights prescribed in paragraph (a) of this Rule, exhibit a high intensity all-round flashing red light.

(d) (i) A power-driven vessel of less than 12 metres in length may in lieu of the lights prescribed in paragraph (a) of this Rule exhibit an all-round white light and sidelights;

(ii) a power-driven vessel of less than 7 metres in length whose maximum speed does not exceed 7 knots may in lieu of the lights prescribed in paragraph (a) of this Rule exhibit an all-round white light and shall, if practicable, also exhibit sidelights;

(iii) the masthead light or all-round white light on a power-driven vessel of less than 12 metres in length may be displaced from the fore and aft centreline of the vessel if centreline fitting is not practicable, provided that the sidelights are combined in one lantern which shall be carried on the fore and aft centreline of the vessel or located as nearly as practicable in the same fore and aft line as the masthead light or the all-round white light.

COMMENT:

Further details are given in Annex I(2) and (3). The second masthead light must be exhibited higher than the forward masthead light under all normal conditions of trim.

Air-cushion vessels

The all-round flashing yellow light must only be exhibited by an air-cushion vessel when operating in the non-displacement mode. The main purpose of this light is to warn other vessels that, as the vessel is in the non-displacement mode, her navigation lights may give a false indication of the direction of travel (see page 92).

WIG craft

The high intensity all-round flashing red light when exhibited by a WIG craft indicates that the WIG craft is taking-off or landing or in flight near the surface and is required by Rule 18(f) to keep well clear of all other vessels and avoid impeding their navigation (see also pages 7–9 and 92).

Submarines

Submarines may also exhibit an amber flashing light as an aid to identification in coastal waters, in addition to the navigation lights of a power-driven vessel. This light is carried above the after 'masthead light', and the forward 'masthead light' of a submarine may be at a lower level than the sidelights.

Small vessels

Power-driven vessels of less than 12 metres in length are permitted to exhibit an all-round white light instead of the masthead light and sternlight, but are required to exhibit sidelights when they are 7 metres or more in length.

Power-driven vessels of less than 7 metres in length are permitted to exhibit an all-round white light, instead of the masthead light, sidelights and sternlight, if their maximum speed does not exceed 7 knots. This provision takes account of the fact that small, low speed vessels may not have sufficient power to exhibit the normal navigation lights. It does not apply to a vessel capable of more than 7 knots which is proceeding at reduced speed.

The all-round light must, apparently, be continuously exhibited when underway at night. It is not sufficient to exhibit it in time to prevent collision, as is permitted for sailing vessels of less than 7 metres. The final sentence of Rule 23(c) requires sidelights to be exhibited if practicable. Rule 21(b) permits any vessel of less than 20 metres in length to combine the sidelights in one lantern. Paragraph (c) was amended in 1981 to extend the use of the all-round white light and to permit it to be displaced from the fore and aft centreline.

RULE 24

Towing and pushing

(a) **A power-driven vessel when towing shall exhibit:**
    (i)   **instead of the light prescribed in Rule 23(a)(i) or (a)(ii), two masthead lights in a vertical line. When the length of the tow, measuring from the stern of the towing vessel to the after end of the tow exceeds 200 metres, three such lights in a vertical line;**
    (ii) **sidelights;**
    (iii) **a sternlight;**
    (iv) **a towing light in a vertical line above the sternlight;**
    (v) **when the length of the two exceeds 200 metres, a diamond shape where it can best be seen.**

(b) **When a pushing vessel and a vessel being pushed ahead are rigidly connected in a composite unit they shall be regarded as a power-driven vessel and exhibit the lights prescribed in Rule 23.**

(c) **A power-driven vessel when pushing ahead or towing alongside, except in the case of a composite unit, shall exhibit:**

   (i) instead of the light prescribed in Rule 23(a)(i) or (a)(ii), two masthead lights in a vertical line;

   (ii) sidelights;

   (iii) a sternlight.

(d) A power-driven vessel to which paragraphs (a) or (c) of this Rule apply shall also comply with Rule 23(a)(ii).

(e) A vessel or object being towed, other than those mentioned in paragraph (g) of this Rule, shall exhibit:

   (i) sidelights;

   (ii) a sternlight;

   (iii) when the length of the tow exceeds 200 metres, a diamond shape where it can best be seen.

(f) Provided that any number of vessels being towed alongside or pushed in a group shall be lighted as one vessel.

   (i) a vessel being pushed ahead, not being part of a composite unit, shall exhibit at the forward end, sidelights;

   (ii) a vessel being towed alongside shall exhibit a sternlight and at the forward end, sidelights.

(g) An inconspicuous, partly submerged vessel or object, or combination of such vessels or objects being towed, shall exhibit:

   (i) if it is less than 25 metres in breadth, one all-round white light at or near the forward end and one at or near the after end except that dracones need not exhibit a light at or near the forward end;

   (ii) if it is 25 metres or more in breadth, two additional all-round white lights at or near the extremities of its breadth;

   (iii) if it exceeds 100 metres in length, additional all-round white lights between the lights prescribed in sub-paragraphs (i) and (ii) so that the distance between the lights shall not exceed 100 metres;

   (iv) a diamond shape at or near the aftermost extremity of the last vessel or object being towed and if the length of the tow exceeds 200 metres an additional diamond shape where it can best be seen and located as far forward as is practicable.

(h) Where from any sufficient cause it is impracticable for a vessel or object being towed to exhibit the lights or shapes prescribed in paragraph (e) or (g) of this Rule, all possible measures shall be taken to light the vessel or object towed or at least to indicate the presence of such vessel or object.

(i) Where from any sufficient cause it is impracticable for a vessel not normally engaged in towing operations to display the lights prescribed by paragraph (a) or (c) of this Rule, such vessel shall not be required to exhibit those lights when engaged in towing another vessel in distress or otherwise in need of assistance. All possible measures shall be taken to indicate the nature of the relationship between the towing vessel and the vessel being towed as authorised by Rule 36 in particular to illuminate the towline.

COMMENT:

Masthead lights

The vertical spacing of the masthead lights prescribed for vessels towing or pushing is specified in Section 2 of Annex I, paragraphs (e) and (i). Paragraphs (a) and (c) of Rule 24 were amended in 1981 to allow the positioning of masthead lights on a mast above or behind the bridge, so as to avoid undue glare. If three masthead lights are to be carried in a vertical line they must be equally spaced, and at distances of not less than 2 metres apart on a vessel of over 20 metres in length.

Towing light

The yellow towing light, to be carried above the sternlight, is only prescribed for a vessel towing another vessel from the stern. The purpose of this light is to enable a vessel towing to be identified when it is being overtaken. A proposal to allow the towing light to be carried above the sternlight of a vessel being towed was not accepted by the 1972 Conference.

Composite unit

Some tugs and barges are capable of being mechanically locked so rigidly that they can operate in the pushing mode as one unit, even on the high seas. They must only show the lights prescribed for a single power-driven vessel.

Towing alongside

A vessel towing alongside is required to show the same lights as a vessel being pushed ahead, i.e., two masthead lights in vertical line, sidelights and a sternlight. A vessel being towed alongside must exhibit a sternlight and sidelights.

After masthead light

The carriage of the 'second masthead light', mentioned in Rule 23(a)(ii), by a vessel towing or pushing ahead is compulsory for vessels over 50 metres in length and optional for vessels of less than 50 metres.

Vessels or objects being towed

Dumb barges and other unmanned vessels or objects being towed must be fitted with the prescribed lights and/or shapes so far as practicable.

Due to the increasing number of tows which consist of inconspicuous or partly submerged objects, a new paragraph (g) has been added, by the amendments of 1981, to improve the provisions for the lighting of such vessels or objects. This new paragraph will apply to flexible oil barges, known as dracones, and to timber floats but may also have application to other vessels or objects being towed.

Vessels not normally engaged in towing operations

A new paragraph (i) has been added, by the amendments of 1981, to make provisions for a vessel not normally engaged in towing operations which assists another vessel by taking it in tow.

Day signal

When the length of the tow, measured from the stern of the towing vessel to the after end of the tow, exceeds 200 metres both the towing vessel and the vessel towed must exhibit the black diamond shape. If the length of tow is less than 200 metres neither vessel shall exhibit the day signal.

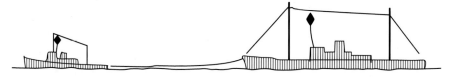

Purpose of special lights

The extra lights prescribed for a vessel towing or pushing ahead are for the purpose of indicating to other vessels that she is not entirely her own mistress and cannot be expected to act in every respect as an ordinary power-driven vessel. In the case of a vessel towing another vessel from the stern they serve to indicate the extra length of obstruction to be expected and the fact that there is a tow-line between the two vessels.

Vessels engaged in towing are only classed as privileged vessels if they are severely restricted in their ability to deviate from their course and are showing the lights or shapes prescribed by Rule 27(b).

RULE 25

Sailing vessels underway and vessels under oars

(a) A sailing vessel underway shall exhibit:
   (i) sidelights;
   (ii) a sternlight.
(b) In a sailing vessel of less than 20 metres in length the lights prescribed in paragraph (a) of this Rule may be combined in one lantern carried at or near the top of the mast where it can best be seen.
(c) A sailing vessel underway may, in addition to the lights prescribed in paragraph (a) of this Rule, exhibit at or near the top of the mast, where they can best be seen, two all-round lights in a vertical line, the upper being red and the lower green, but these lights shall not be exhibited in conjunction with the combined lantern permitted by paragraph (b) of this Rule.

(d) (i) A sailing vessel of less than 7 metres in length shall, if practicable, exhibit the lights prescribed in paragraph (a) or (b) of this Rule, but if she does not, she shall have ready at hand an electric torch or lighted lantern showing a white light which shall be exhibited in sufficient time to prevent collision.

(ii) A vessel under oars may exhibit the lights prescribed in this Rule for sailing vessels, but if she does not, she shall have ready at hand an electric torch or lighted lantern showing a white light which shall be exhibited in sufficient time to prevent collision.

(e) A vessel proceeding under sail when also being propelled by machinery shall exhibit forward where it can best be seen a conical shape, apex downwards.

COMMENT:

Combined light

Paragraph (b) was amended in 1981 to extend the provision permitting sailing vessels to combine the sidelights and sternlight in one lantern from vessels of less than 12 metres to vessels of less than 20 metres in length. As this light is to be carried at or near the top of the mast it is less likely to be obscured by the sails than the sidelights but it may be less effectively screened. The screening requirements are specified in Section 9 of Annex I. The combined lantern is expected to become popular with yachtsmen.

All-round red and green lights

Sailing vessels are permitted to exhibit red and green lights in a vertical line, visible all-round the horizon. These lights are optional for all sailing vessels. They must not be exhibited in conjunction with the combined lantern, as this may cause confusion.

Small sailing vessels

The sidelights of sailing vessels of less than 20 metres in length may be combined in one lantern in accordance with Rule 21(b). If practicable, sailing vessels of less than 7 metres in length are required to exhibit either the sidelights and sternlight or a combined lantern. The showing of a torch or white light in sufficient time to prevent collision is not specified as an alternative for *any* sailing vessel of less than 7 metres. Such an arrangement would be a poor substitute for continuously exhibited sidelights and sternlight which enable a sailing vessel to be identified in a crossing situation and give some indication of aspect.

Vessels under oars

Vessels under oars are given the option of either exhibiting sidelights and sternlight, which could be combined in one lantern, or showing an electric torch or white light in sufficient time to prevent collision.

Under sail and power

A vessel under sail and power is required to carry a black conical shape, point downwards. In Section 6 of Annex I it is specified that such a cone shall have a height and base diameter of not less than 0.6 metre but that vessels of less than 20 metres in length may use a smaller cone, commensurate with the size of the vessel.

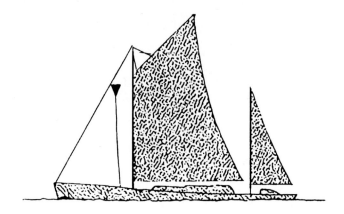

RULE 26

Fishing vessels

(a) A vessel engaged in fishing, whether underway or at anchor, shall exhibit only the lights and shapes prescribed in this Rule.

(b) A vessel when engaged in trawling, by which is meant the dragging through the water of a dredge net or other apparatus used as a fishing appliance, shall exhibit:

   (i) two all-round lights in a vertical line, the upper being green and the lower white, or a shape consisting of two cones with their apexes together in a vertical line one above the other;

   (ii) a masthead light abaft of and higher than the all-round green light; a vessel of less than 50 metres in length shall not be obliged to exhibit such a light but may do so;

   (iii) when making way through the water, in addition to the lights prescribed in this paragraph, sidelights and a sternlight.

(c) A vessel engaged in fishing, other than trawling, shall exhibit:

   (i) two all-round lights in a vertical line, the upper being red and the lower white, or a shape consisting of two cones with apexes together in a vertical line one above the other;

   (ii) when there is outlying gear extending more than 150 metres horizontally from the vessel, an all-round white light or a cone apex upwards in the direction of the gear;

(iii) **when making way through the water, in addition to the lights prescribed in this paragraph, sidelights and a sternlight.**
(d) **The additional signals described in Annex II to these Regulations apply to a vessel engaged in fishing in close proximity to other vessels engaged in fishing.**
(e) **A vessel when not engaged in fishing shall not exhibit the lights or shapes prescribed in this Rule, but only those prescribed for a vessel of her length.**

COMMENT:

Vessels engaged in fishing

Any vessel engaged in fishing must exhibit the lights and shapes prescribed by Rule 26; there is no relaxation for small fishing vessels. The definition of 'vessel engaged in fishing' is given in Rule 3(d) (see pages 5–6). A vessel shooting or hauling nets is considered to be engaged in fishing. A vessel engaged in fishing when at anchor, or with nets fast to a rock, must show only the lights prescribed in this Rule and may not exhibit the anchor lights prescribed in Rule 30.

Vessels engaged in fishing cannot be regarded as 'not under command' or 'restricted in their ability to manoeuvre'. Even if their engines or steering gear becomes defective they must show only the lights and shapes prescribed by Rule 26. However, such vessels are given a high degree of privilege and are only required by Rule 18 to keep clear of other hampered vessels 'so far as possible'.

Fishing vessels not engaged in fishing must exhibit the appropriate lights for vessels of their length and must not show the lights prescribed in Rule 26.

Making way

All vessels engaged in fishing are required to show sidelights and a sternlight when making way through the water but must not show these lights when under way and stopped. A fishing vessel of less than 20 metres in length may have the sidelights combined in one lantern.

A single white light in conjunction with a red light over a white light in vertical line could indicate a vessel engaged in fishing, other than trawling, either making way and being overtaken or not making way with gear extending more than 150 metres in the direction of the single white light.

Masthead light when trawling

A masthead light must be exhibited by a vessel over 50 metres in length engaged in trawling and is optional for smaller vessels. It must be carried abaft, higher than the all-round green light and be visible for at least 6 miles. The second masthead light specified in Rule 2(a)(ii) can be used for this purpose.

Day signals

In 1993 the eighteenth Assembly of IMO approved amendments to sub-paragraphs (b)(i) and (c)(i) to delete the option of displaying a basket in lieu of two cones with their apexes together, one above the other, by fishing vessels of less than 20 metres in length. It was considered that in practice a basket was not a satisfactory substitute for the two cones.

Vessels fishing in close proximity

Fishing vessels of 20 metres or more in length when engaged in trawling, including pair trawling, in close proximity to other vessels are required to exhibit the additional signals prescribed in Annex II. Vessels engaged in trawling, of less than 20 metres in length, and vessels of any length engaged in fishing with purse seine gear may exhibit the signals prescribed in Annex II but are not obliged to do so. Rule 26(d) and Annex II Section 2 were amended by the eighteenth Assembly of IMO to give more emphasis to the provisions of Annex II and to make the signals mandatory for vessels of 20 metres or more in length trawling in close proximity.

Annex II refers specifically to vessels engaged in trawling or using purse seine gear but Rule 1(c) permits the Government of any State to make special rules relating to signal lights for fishing vessels engaged in fishing as a fleet which could apply to other types of fishing. The additional lights referred to in Annex II must be carried at a lower level, and be visible at a shorter range, than the all-round lights prescribed in Rule 26(b)(i) and (c)(i).

Signals to attract attention

If necessary to attract attention of another vessel, any vessel, including a vessel engaged in fishing, may use light or sound signals and may direct the beam of a searchlight in the direction of a danger in accordance with Rule 36.

RULE 27

Vessels not under command or restricted in their ability to manœuvre

(a) A vessel not under command shall exhibit:
  (i) two all-round red lights in a vertical line where they can best be seen;
  (ii) two balls or similar shapes in a vertical line where they can best be seen;
  (iii) when making way through the water, in addition to the lights prescribed in this paragraph, sidelights and a sternlight.
(b) A vessel restricted in her ability to manœuvre, except a vessel engaged in mine clearance operations, shall exhibit:
  (i) three all-round lights in a vertical line where they can best be seen. The highest and lowest of these lights shall be red and the middle light shall be white;
  (ii) three shapes in a vertical line where they can best be seen. The highest and lowest of these shapes shall be balls and the middle one a diamond;
  (iii) when making way through the water, masthead light or lights, sidelights and a sternlight, in addition to the lights prescribed in sub-paragraph (i);

(iv) when at anchor, in addition to the lights or shapes prescribed in sub-paragraphs (i) and (ii), the light, lights or shape prescribed in Rule 30.

(c) A power-driven vessel engaged in a towing operation such as severely restricts the towing vessel and her tow in their ability to deviate from their course shall, in addition to the lights or shapes prescribed in Rule 24(a), exhibit the lights or shapes prescribed in sub-paragraphs (b)(i) and (ii) of this Rule.

(d) A vessel engaged in dredging or underwater operations, when restricted in her ability to manœuvre, shall exhibit the lights and shapes prescribed in sub-paragraphs (b)(i), (ii) and (iii) of this Rule and shall in addition, when an obstruction exists, exhibit:

    (i)   two all-round red lights or two balls in a vertical line to indicate the side on which the obstruction exists;

    (ii) two all-round green lights or two diamonds in a vertical line to indicate the side on which another vessel may pass;

    (iii) when at anchor, the lights or shapes prescribed in this paragraph instead of the lights or shapes prescribed in Rule 30.

(e) Whenever the size of a vessel engaged in diving operations makes it impracticable to exhibit all lights and shapes prescribed in paragraph (d) of this Rule, the following shall be exhibited:

    (i)   three all-round lights in a vertical line where they can best be seen. The highest and lowest of these lights shall be red and the middle light shall be white;

    (ii) a rigid replica of the International Code flag 'A' not less than 1 metre in height. Measures shall be taken to ensure its all-round visibility.

(f) A vessel engaged in mine clearance operations shall in addition to the lights prescribed for a power-driven vessel in Rule 23 or for a vessel at anchor in Rule 30 as appropriate, exhibit three all-round green lights or three balls. One of these lights or shapes shall be exhibited near the foremast head and one at each end of the fore yard. These lights or shapes indicate that it is dangerous for another vessel to approach within 1000 metres of the mine clearance vessel.

(g) Vessels of less than 12 metres in length, except those engaged in diving operations, shall not be required to exhibit the lights and shapes prescribed in this Rule.

(h) The signals prescribed in this Rule are not signals of vessels in distress and requiring assistance. Such signals are contained in Annex IV to these Regulations.

COMMENT:

Vessel not under command

The definition of a vessel not under command is given in Rule 3(f) (see pages 5–6). It is, of course, implied that the vessel must be under way. A power-driven vessel or sailing vessel which through some exceptional circumstance is unable to manœuvre as required by these Rules should show only the lights and shapes prescribed in Rule 27(a).

Vessel restricted in her ability to manœuvre

The definition of 'vessel restricted in her ability to manœuvre' and the circumstances under which a vessel should be considered to be in this category are given in Rule 3(g) (see pages 5–6).

In 2008, the Maritime Safety Committee of IMO (MSC./84) approved MSC.1/ Circ.1260 and adopted the following unified interpretation related to Rule 27(a)(i) and (b)(i):

"Not under command" (NUC) all-round red lights (Rule 27(a)(i)), may be used as part of the "Restricted Ability to Manœuvre" (RAM) lights (Rule 27(b)(i)), provided the vertical and horizontal distances, required by COLREG 1972, are complied with and the electrical system is arranged, so that the all-round white light (RAM) may be switched on independently from the two all-round red lights (NUC).

Special operations at anchor

Rule 27(b)(iv) requires a vessel restricted in her ability to manœuvre, except a vessel engaged in mine clearance operations, when at anchor, to show the three all-round lights and three shapes in vertical line in addition to anchor signals. However, Rule 27(d)(iii) requires a vessel engaged in dredging or underwater operations, when restricted in her ability to manœuvre, to show the lights and shapes specified in Rule 27(d)(i) and (ii) instead of the anchor signals.

Vessels engaged in transferring persons, provisions or cargo when at anchor are not included in the category referred to in Rule 3(g)(iii) which applies only to vessels under way. There is normally no necessity for such vessels to warn other vessels to keep well clear.

Vessel engaged in a difficult towing operation

Any power-driven vessel engaged in a towing operation such as severely restricts both the towing vessel and her tow in their ability to deviate from their course is required to show the lights and shapes prescribed in Rule 27(b)(i) and (ii).

Paragraph (c) of Rule 27 was amended in 1981 to bring its wording in line with Rule 3(g)(vi) and to remove any ambiguity on its application to the vessel or object being towed. The lights and shapes prescribed in Rule 27(b)(i) and (ii) may not be exhibited by the vessel or object being towed.

Some vessels engaged in a difficult towing operation may be justified in exhibiting the lights and shapes specified in Rule 27(b) for the duration of the tow but others may only become severely restricted in their ability to deviate from their course for a relatively short period as a result of rough weather.

Vessel engaged in dredging or underwater operations

Vessels engaged in underwater operations should only show the lights and shapes prescribed in Rule 27(b) if they are severely restricted in their ability to manœuvre. The additional lights and shapes specified in Rule 27(d) should only be exhibited if there is an obstruction on one side and the other side is clear. Paragraph (d) was amended in 1981 to clarify its text. The positions at which these additional lights should be carried are specified in Annex I.4(b).

A vessel engaged in surveying may show the International Code signals IR or PO. A seismic survey vessel shooting lines may display flag B, or at night a single red light in addition to the prescribed lights. Further details are given in the Mariner's Handbook which advises other vessels to give survey vessels a wide berth (preferably more than 2 miles).

Paragraph (e) of Rule 27 as amended in 1981 prescribes new provisions for small vessels engaged in diving operations which find it impracticable to exhibit the lights or shapes required in paragraph (d).

Vessels engaged in mine-clearance operations

The three all-round green lights prescribed in Rule 27 (f) must be shown in addition to the lights prescribed for a power-driven vessel. Minesweeping vessels were previously exempted from carrying a second masthead light but this no longer applies. However, Rule 1(e) permits Governments to make separate provisions for vessels of special construction.

Rule 27(f) was amended in 1981 to extend its application to all mine clearance operations.

Small hampered vessels

Rule 27(g) was amended in 1981. The exemption from having NUC lights and shapes in readiness was extended from vessels of less than 7 metres in length to vessels of less than 12 metres in length and the Rule, which originally referred to lights only, now refers to both lights and shapes.

RULE 28

Vessels constrained by their draught

A vessel constrained by her draught may, in addition to the lights prescribed for power-driven vessels in Rule 23, exhibit where they can best be seen three all-round red lights in a vertical line, or a cylinder.

COMMENT:

A vessel constrained by her draught is not required to exhibit the three all-round lights in a vertical line, or the cylinder, as it is sometimes difficult to know whether the restriction in ability to alter course is sufficient to justify their use. If the signals are not shown the privilege extended to such vessels by Rule 18 would not, of course, apply. The three red lights must be shown in conjunction with the lights for a power-driven vessel under way.

RULE 29

Pilot vessels

(a) A vessel engaged in pilotage duty shall exhibit:
   (i) at or near the masthead, two all-round lights in a vertical line, the upper being white and the lower red;
   (ii) when underway, in addition, sidelights and a sternlight;
   (iii) when at anchor, in addition to the lights prescribed in subparagraph (i), the light, lights or shape prescribed in Rule 30 for vessels at anchor.
(b) A pilot vessel when not engaged on pilotage duty shall exhibit the lights or shapes prescribed for a similar vessel of her length.

COMMENT:

If necessary to attract the attention of another vessel, a pilot vessel may make light or sound signals that cannot be mistaken for any signal authorised elsewhere in the Rules, in accordance with Rule 36.

Rule 29(a)(iii) was amended in 1981 for purposes of clarification.

RULE 30

Anchored vessels and vessels aground

(a) A vessel at anchor shall exhibit where it can best be seen:
   (i) in the fore part, an all-round white light or one ball;
   (ii) at or near the stern and at a lower level than the light prescribed in sub-paragraph (i), an all-round white light.
(b) A vessel of less than 50 metres in length may exhibit an all-round white light where it can best be seen instead of the lights prescribed in paragraph (a) of this Rule.
(c) A vessel at anchor may, and a vessel of 100 metres and more in length shall, also use the available working or equivalent lights to illuminate her decks.
(d) A vessel aground shall exhibit the lights prescribed in paragraph (a) or (b) of this Rule and in addition, where they can best be seen:
   (i) two all-round red lights in a vertical line;
   (ii) three balls in a vertical line.
(e) A vessel of less than 7 metres in length, when at anchor, not in or near a narrow channel, fairway or anchorage, or where other vessels normally navigate,

**shall not be required to exhibit the lights or shapes prescribed in paragraphs
(a) and (b) of this Rule.**

**(f) A vessel of less than 12 metres in length, when aground, shall not be
required to exhibit the lights or shapes prescribed in sub-paragraphs
(d)(i) and (ii) of this Rule.**

COMMENT:

At anchor

Vessels have been held by the Courts to be at anchor so long as the anchor is down and
is holding. When weighing anchor a vessel continues to be at anchor until the anchor
is out of the ground.

Forest Lake–Janet Quinn

*I have consulted the Elder Brethren as to when the right moment is to drop the anchor
ball, and they advise me, and I accept, that it should be dropped as soon as the anchor
is clear of the sea floor. (Mr Justice Karminski, 1967)*
    A vessel moored to buoys may be considered to be at anchor. A vessel dredging
down with the tide or dragging her anchor is under way, but if the anchor fouls an
obstruction and is held she is at anchor.

Position of anchor signals

A vessel of less than 50 metres in length is permitted to exhibit a single all-round
white light where it can best be seen, not necessarily in the fore part. Any vessel
which exhibits two anchor lights, including a vessel of less than 50 metres, is required
to exhibit the forward one in the fore part of the vessel. The ball must always be
exhibited in the fore part.
    Further details concerning the positions of anchor lights are given in Annex I(2)(k).

Working lights

Paragraph (c) contains a provision, based on common practice, requiring vessels of
100 metres or more in length to use lights to illuminate her decks, and permitting such
lights to be used by smaller vessels. These lights must be switched off as soon as the
vessel gets under way.

Small vessels

Rule 30(e) was amended in 1981 to limit its application to vessels at anchor, while for
vessels of less than 12 metres in length and aground an additional paragraph (f) is
included to exempt such vessels from exhibiting the lights and shapes prescribed in
Rule 30(d)(i) and (ii). However, the anchor signals prescribed in Rule 30(a) and (b)
must be exhibited.

RULE 31

Seaplanes and WIG craft

Where it is impracticable for a seaplane or a WIG craft to exhibit lights and shapes of the characteristics or in the positions prescribed in the Rules of this Part she shall exhibit lights and shapes as closely similar in characteristics and position as is possible.

COMMENT:

Seaplanes and WIG craft will normally have a 'masthead light' in the forepart and sidelights on the wing tips when under way on the water. A large seaplane at anchor may have white lights on the wing tips in addition to white lights forward and aft.

Part D – Sound and light signals

RULE 32

Definitions

(a) The word 'whistle' means any sound signalling appliance capable of producing the prescribed blasts and which complies with the specifications in Annex III to these Regulations.

(b) The term 'short blast' means a blast of about one second's duration.

(c) The term 'prolonged blast' means a blast of from four to six seconds' duration.

COMMENT:

All whistle signals prescribed in the Rules are specified in terms of short blasts and prolonged blasts.

RULE 33

Equipment for sound signals

(a) A vessel of 12 metres or more in length shall be provided with a whistle, a vessel of 20 metres or more in length shall be provided with a bell in addition to a whistle, and a vessel of 100 metres or more in length shall, in addition, be provided with a gong, the tone and sound of which cannot be confused with that of the bell. The whistle, bell and gong shall comply with the specifications in Annex III to these Regulations. The bell or gong or both may be replaced by other equipment having the same respective sound characteristics, provided that manual sounding of the prescribed signals shall always be possible.

(b) A vessel of less than 12 metres in length shall not be obliged to carry the sound signalling appliances prescribed in paragraph (a) of this Rule but if she does not, she shall be provided with some other means of making an efficient sound signal.

COMMENT:

Rule 33(a) was amended by the 22nd IMO Assembly, by which ships of less than 20 metres in length no longer are required to be provided with a bell.

A Guide to the Collision Avoidance Rules. DOI: 10.1016/B978-0-08-097170-4.00004-0

Whistle frequencies

The specifications for a whistle are given in Section 1 of Annex III. The whistle frequencies are related to the length of vessel to ensure a wide variety of characteristics. The fundamental frequency of the whistle for a vessel 200 metres or more in length must be between 70 and 200 Hz to give a relatively deep tone. Vessels of less than 75 metres are required to have a whistle frequency of between 250 and 700 Hz giving a relatively shrill tone. The whistle of a vessel of intermediate size must have a frequency in the range of 130–350 Hz.

Range of audibility

The Annex does not give a statutory minimum range of audibility for a whistle in still conditions. Some typical audibility ranges are given for the whistles of vessels of different size with a warning that the range of audibility is extremely variable and depends critically on weather conditions. A typical audibility range of 2 miles in still conditions is quoted for vessels of 200 metres or more in length.

Bell and gong

Section 2 of Annex III gives technical details of the bell and gong. Rule 33(a) permits the bell or gong to be replaced by some other equipment having the same respective sound characteristics so that automatic devices can be used. However, manual sounding of bell and gong signals must always be possible.

Small vessels

Small vessels not carrying the specified appliances must be provided with alternative means of making an effective sound signal such as an aerosol type foghorn.

RULE 34

Manœuvring and warning signals

(a) When vessels are in sight of one another, a power-driven vessel underway, when manœuvring as authorised or required by these Rules, shall indicate that manœuvre by the following signals on her whistle:
   – one short blast to mean 'I am altering my course to starboard';
   – two short blasts to mean 'I am altering my course to port';
   – three short blasts to mean 'I am operating astern propulsion'.
(b) Any vessel may supplement the whistle signals prescribed in paragraph (a) of this Rule by light signals, repeated as appropriate, while the manœuvre is being carried out:
   (i)   these light signals shall have the following significance:
      – one flash to mean 'I am altering my course to starboard';

- two flashes to mean 'I am altering my course to port';
- three flashes to mean 'I am operating astern propulsion';
(ii) the duration of each flash shall be about one second, the interval between flashes shall be about one second, and the interval between successive signals shall be not less than ten seconds;
(iii) the light used for this signal shall, if fitted, be an all-round white light, visible at a minimum range of 5 miles, and shall comply with the provisions of Annex I to these Regulations.
(c) When in sight of one another in a narrow channel or fairway:
(i) a vessel intending to overtake another shall in compliance with Rule 9(e)(i) indicate her intention by the following signals on her whistle:
- two prolonged blasts followed by one short blast to mean 'I intend to overtake you on your starboard side';
- two prolonged blasts followed by two short blasts to mean 'I intend to overtake you on your port side';
(ii) the vessel about to be overtaken when acting in accordance with Rule 9(e)(i) shall indicate her agreement by the following signal on her whistle:
- one prolonged, one short, one prolonged and one short blast, in that order.
(d) When vessels in sight of one another are approaching each other and from any cause either vessel fails to understand the intentions or actions of the other, or is in doubt whether sufficient action is being taken by the other to avoid collision, the vessel in doubt shall immediately indicate such doubt by giving at least five short and rapid blasts on the whistle. Such signal may be supplemented by a light signal of at least five short and rapid flashes.
(e) A vessel nearing a bend or an area of a channel or fairway where other vessels may be obscured by an intervening obstruction shall sound one prolonged blast. Such signal shall be answered with a prolonged blast by any approaching vessel that may be within hearing around the bend or behind the intervening obstruction.
(f) If whistles are fitted on a vessel at a distance apart of more than 100 metres, one whistle only shall be used for giving manœuvring and warning signals.

COMMENT:

In sight of one another

The signals described in paragraphs (a), (b), (c) and (d) are only to be given by vessels in visual sight of one another and paragraph (e) is obviously intended to apply in clear visibility. Manœuvring signals should not be given when avoiding action is taken at close range for a vessel detected by radar but not visually sighted. However, a vessel is unlikely to be excused for not sounding the signals if failure to sight the other vessel is due to a bad visual look-out.

Lucille Bloomfield–Ronda

'In sight', in my view, means something which is visible if you take the trouble to keep a look-out, and that of course is the position here. In short, the obligation to make sound signals is not excused by the fact that nobody looks to see what there is about. (Mr Justice Karminski, 1966).

Application to sailing vessels

Rule 34(a) refers only to power-driven vessels. Sailing vessels are not required to give the manœuvring signals when taking action to avoid collision. The remaining paragraphs of Rule 34 apply to all vessels. In particular, it should be noted that sailing vessels are now required to give the signal of at least five short and rapid blasts when in doubt about the intentions or actions of the other vessel.

Authorised or required

The whistle signals prescribed in Rule 34(a) must be given when manœuvring as authorised or required by these Rules. Signals to indicate course alterations are not required if the helm is used to counteract the effect of the tide, or to check the swing of the vessel when moving astern. A vessel which put her engines astern while turning in a river, without coming bodily astern, was held to have been under no obligation to sound three short blasts.

Even a small alteration of course must, generally, be indicated by the appropriate whistle signal if it is authorised or required by the Rules.

Varmdo–Jeanne M

... If the helm action be light it is perhaps even more important to give notice thereof by whistle signal, since it is clearly less easy for the vessel to perceive the effects of such helm action than when the action is sudden and heavy. (Mr Justice Langton, 1939)

However, in the *Royalgate–Peter* (1967) it was held that there was no necessity for sound signals to be given by a vessel which altered course 5° to port, then came back to her original course about 5 minutes later, as this was not really changing course in the ordinary sense.

The word 'authorised' would cover action not specifically required by the Rules such as that made as a necessary departure to avoid immediate danger, or as a precaution required by the ordinary practice of seamen, in accordance with Rule 2.

Sound signals need not be given if action is taken for a vessel in sight, at long range, before risk of collision begins to exist, but if the Rules do apply the signals must be sounded even if it is thought that they may not be heard.

Haugland–Karamea

The chief officer of the Haugland, when asked why he did not give the signal, gave as his reason: 'Because it appeared to me that the Karamea was too far away; she would not hear it'. This was a clear infraction of the Rule. If the vessels are in sight the signals must be given. The obligation is not conditional upon the signal being audible to the other vessel. It is easy to understand why the rule was drawn in these peremptory terms. It would be very dangerous if the officer in charge were encouraged to speculate as to whether the signal, if given, would be heard; he must give it if in sight. (Viscount Finlay, House of Lords, 1921)

Fremona–Electra

We know perfectly well that the officers of the watch on occasions like this sometimes think it will disturb the ship – the master or somebody – if they blow the whistle and it does not turn out to be necessary. We cannot accept that as any excuse. The Rule is perfectly clear. The word 'shall' is there – shall sound his whistle – and that word 'shall' must be obeyed; and if officers of the watch for any reason choose to neglect the duty which that Rule imposes upon them they have only themselves to blame if they are found in fault. (Mr Justice Bargrave Deane, 1907)

Signals for action which is not authorised

When a power-driven vessel which is in sight of, and within hearing distance of, another vessel, takes action which is not authorised or required by the Rules, it may in certain circumstances be good seamanship to give the signals prescribed in Rule 34(a).

Operating astern propulsion

A signal of three short blasts does not necessarily mean that the vessel sounding it is moving astern through the water. It may take several minutes to check the headway with astern propulsion. The term 'I am operating astern propulsion' has been substituted for 'my engines are going astern', used in the previous Regulations, as some vessels do not have to reverse their engines to achieve astern propulsion.

Visual signal

In Annex I(12) it is specified that the manœuvring light shall be carried, where practicable, at a minimum height of 2 metres vertically above the forward masthead light. This should ensure that the light, if fitted, will be conspicuous.

The noise level on some ships, particularly motor ships, is often very high making it difficult for sound signals to be heard. The visual signal, especially when repeated while the manœuvre is being carried out, gives an important additional indication of action taken to avoid collision. As the signal is not compulsory it need not be used in circumstances when it is likely to confuse other vessels but it may sometimes prove to be invaluable. It is to be hoped that many vessels will be provided with this new manœuvring light.

Signals for overtaking in a narrow channel

Paragraph (c) specifies the sound signals to be used by vessels acting in compliance with Rule 9(e) (see page 49). No signal is prescribed when the vessel about to be overtaken is not in agreement that it is safe to pass, but Rule 9(e) states that, if in doubt, such a vessel may give the signal of at least five short blasts prescribed in Rule 34(d). This signal can be used to acknowledge that the signals of the vessel intending to overtake have been heard as well as to indicate doubt as to the wisdom of attempting to pass in that part of the channel. The overtaking vessel must repeat the signals and receive a signal of agreement before attempting to pass. Communication by VHF radiotelephone would be useful in such circumstances.

Note: The white masthead lights should be placed above and clear of all other lights.

Wake-up signal

A give-way vessel is required to take early and substantial action to keep well clear by Rule 16, and Rule 8 also requires action to avoid collision to be positive and to be made in ample time. If the give-way vessel fails to take positive early action the stand-on vessel is obliged to give at least five short and rapid blasts on the whistle. It should be noted that the signal is to be at least five short blasts; if there is no quick response the sequence should be continued or the signal repeated in the hope of attracting attention.

The Rule now specifically refers to the use of a light signal of at least five short and rapid flashes to supplement the whistle signal. This signal can be made with a signalling lamp and has been in common use. The light signal may be more effective than the whistle, especially at night.

The signal prescribed in Rule 34(d) must be used by any vessel which fails to understand the intentions or actions of the other vessel. It is specifically referred to in Rule 9(d) and (e) for use in narrow channels.

RULE 35

Sound signals in restricted visibility

In or near an area of restricted visibility, whether by day or night, the signals prescribed in this Rule shall be used as follows:

(a) A power-driven vessel making way through the water shall sound at intervals of not more than 2 minutes one prolonged blast.

(b) A power-driven vessel underway but stopped and making no way through the water shall sound at intervals of not more than 2 minutes two prolonged blasts in succession with an interval of about 2 seconds between them.

(c) A vessel not under command, a vessel restricted in her ability to manœuvre, a vessel constrained by her draught, a sailing vessel, a vessel engaged in

fishing and a vessel engaged in towing or pushing another vessel shall, instead of the signals prescribed in paragraphs (a) or (b) of this Rule, sound at intervals of not more than 2 minutes three blasts in succession, namely one prolonged followed by two short blasts.

(d) A vessel engaged in fishing, when at anchor, and a vessel restricted in her ability to manœuvre when carrying out her work at anchor, shall instead of the signals prescribed in paragraph (g) of this Rule sound the signal prescribed in paragraph (c) of the Rule.

(e) A vessel towed or if more than one vessel is towed the last vessel of the tow, if manned, shall at intervals of not more than 2 minutes sound four blasts in succession, namely one prolonged followed by three short blasts. When practicable, this signal shall be made immediately after the signal made by the towing vessel.

(f) When a pushing vessel and a vessel being pushed ahead are rigidly connected in a composite unit they shall be regarded as a power-driven vessel and shall give the signals prescribed in paragraphs (a) or (b) of this Rule.

(g) A vessel at anchor shall at intervals of not more than one minute ring the bell rapidly for about 5 seconds. In a vessel of 100 metres or more in length the bell shall be sounded in the forepart of the vessel and immediately after the ringing of the bell the gong shall be sounded rapidly for about 5 seconds in the after part of the vessel. A vessel at anchor may in addition sound three blasts in succession, namely one short, one prolonged and one short blast, to give warning of her position and of the possibility of collision to an approaching vessel.

(h) A vessel aground shall give the bell signal and if required the gong signal prescribed in paragraph (g) of this Rule and shall, in addition, give three separate and distinct strokes on the bell immediately before and after the rapid ringing of the bell. A vessel aground may in addition sound an appropriate whistle signal.

(i) A vessel of 12 metres or more but less than 20 metres in length shall not be obliged to give the bell signals prescribed in paragraphs (g) and (h) of this Rule. However, if she does not, she shall make some other efficient sound signal at intervals of not more than 2 minutes.

(j) A vessel of less than 12 metres in length shall not be obliged to give the above-mentioned signals but, if she does not, shall make some other efficient sound signal at intervals of not more than 2 minutes.

(k) A pilot vessel when engaged in pilotage duty may in addition to the signals prescribed in paragraphs (a), (b) or (g) of this Rule sound an identity signal consisting of four short blasts.

COMMENT:

New paragraph (i) was added to Rule 35 in 2001 as a consequence to the amendment to Rule 33(a).

In or near an area of restricted visibility

Fog signals must also be given when navigating near an area of restricted visibility, especially when approaching such an area. Rule 19, which relates to conduct in fog, applies when in or near an area of restricted visibility (see page 93).

The density of fog which would necessitate the use of fog signals has not been defined. There would be little point in sounding signals if the visibility is greater than the range of audibility of the appliance being used. However, it would be prudent to consider an upper limit of visibility greater than the audibility ranges quoted in Annex III as the equipment may be capable of being heard at a greater distance than the typical ranges given and it is difficult to determine the exact extent of the visibility.

Intervals between whistle signals

All fog signals to be given by whistle are to be sounded at intervals of not more than two minutes. Some whistle signals were previously required to be given at intervals of not more than one minute but at the 1972 Conference it was decided to standardise the maximum interval for all whistle signals at two minutes as temporary deafness can be caused by sounding the whistle too frequently. Bell and gong signals must still be sounded at intervals of not more than one minute.

The Rule specifies the maximum interval between signals. When other vessels are known to be close by the whistle signals should be given at intervals of less than two minutes. Sounding the signals more frequently will usually give other vessels, which may not have operational radar, greater opportunity to assess the approximate bearing.

Two prolonged blasts

The two blast signal should not be given until it is certain that the vessel has ceased making way through the water.

Lifland–Rosa Luxembourg

I think it is most important that that distinction should be appreciated and observed and that a vessel should not be heard to say: 'Oh, well, if I was not quite stopped I was very nearly stopped and you must not be hard on me, because it is very difficult to tell when one is stopped.' You are not to blow this signal until you are stopped, and you must be quite certain that you really are. (Mr Justice Langton, 1934)

Sailing vessels

Sailing vessels are required to sound the signals prescribed in Rule 35(c) for hampered vessels. The Conference decided not to retain the signals indicating how a vessel was sailing in relation to the wind as this information is not usually of much value to other vessels and the previous signals of one, two or three blasts could be confused with manœuvring signals.

The signal prescribed by Rule 35(c) must only be given by a vessel which is under sail and under way. When at anchor a yacht or sailing ship must sound the signal prescribed in Rule 35(g).

Vessels engaged in towing

The sound signal prescribed in Rule 35(c) is to be sounded by almost all categories of vessel given some degree of privilege by Rule 18 but is not restricted to vessels towing which are engaged in a difficult towing operation. Any vessel engaged in towing must give the signal of a prolonged blast followed by two short blasts.

Any vessel being towed is required to give the separate sound signal prescribed in Rule 35(e) but is only obliged to give the signal if she is manned. However, it may be considered as a precaution required by Rule 2(a) that arrangements be made for such a signal to be given, especially in the case of a long tow, as the towing vessel cannot be identified as such by her fog signal.

A tug fast to a vessel, but not towing, should not give the fog signals for a towing vessel. When one such case came before the Courts it was held that the vessel attached to the tug should have given the signals for a vessel under way unaccompanied by any signals from the tug.

A vessel pushing another vessel

A vessel pushing another vessel must sound the same signals as a vessel towing. Rule 35(f) requires vessels rigidly connected in a composite unit to give the signals of a power-driven vessel.

Vessel at anchor

A vessel of 100 metres or more in length at anchor is required to sound a gong in the after part immediately after sounding the bell.

Annex III does not give the typical audibility ranges of the bell and gong, but these are likely to be relatively low. When lying at anchor in congested waters, or when another vessel seems to be approaching too closely, the stronger whistle signal permitted by Rule 35(g) should be sounded.

Special operations or fishing at anchor

Paragraph (d) is a new paragraph, which was inserted when the Rules were amended in 1981, to remove any uncertainty about the fog signal to be sounded by a vessel engaged in fishing or special operations when at anchor. Such vessels are required to sound the whistle signal prescribed in Rule 35(c) to warn other vessels to keep well clear.

A vessel engaged in replenishment or transferring persons, provisions or cargo is, according to Rule 3(g)(iii), only to be regarded as 'a vessel restricted in her ability to manoeuvre' when she is under way. A vessel engaged in such operations when at anchor in restricted visibility should sound the signal prescribed in Rule 35(g).

Vessel aground

The sound signals for a vessel aground are the same as those prescribed in Rule 15(c)(vii) of the 1960 Regulations. A vessel aground of 100 metres or more in length should sound the gong immediately after sounding the second set of three strokes on the bell.

A new provision is that a vessel aground is permitted to sound an appropriate whistle signal. The character of the signal is not given as the Conference could not decide on a signal which would be suitable for all circumstances. The signal 'U' (two short blasts followed by one prolonged blast), meaning 'you are running into danger', would usually be appropriate for the purpose of warning other vessels to keep well clear.

Pilot vessels

Every pilot vessel, including a sailing pilot vessel, may sound the identity signal consisting of four short blasts. Some vessels engaged in pilotage duty are permitted by local rules to give an alternative form of identity signal. A vessel using the identity signal must continue to sound the fog signal at the prescribed intervals.

RULE 36

Signals to attract attention

If necessary to attract the attention of another vessel any vessel may make light or sound signals that cannot be mistaken for any signal authorised elsewhere in these Rules, or may direct the beam of her searchlight in the direction of the danger, in such a way as not to embarrass any vessel. Any light to attract the attention of another vessel shall be such that it cannot be mistaken for any aid to navigation. For the purpose of this Rule the use of high intensity intermittent or revolving lights, such as strobe lights, shall be avoided.

COMMENT:

Any signal may be used which cannot be mistaken for a signal authorised elsewhere in the Rules, including a flare-up light. A sailing vessel could use a torch or searchlight to illuminate the sails.

Light or sound signals which could be mistaken for signals authorised elsewhere in the Rules must not be used to attract the attention of another vessel. In particular, signals which could be confused with those authorised by Rule 37 are not to be used unless the vessel is in distress. A very long blast on the whistle could, for instance, be taken to be 'a continuous sounding with any fog-signalling apparatus' (Annex IV1(b)).

The last two sentences of Rule 36 were added with the amendments of 1981.

RULE 37

Distress signals

When a vessel is in distress and requires assistance she shall use or exhibit the signals described in Annex IV to these Regulations.

COMMENT:

At the 1972 Conference several countries proposed that the distress signals be deleted from the Regulations as they have nothing to do with the prevention of collisions at sea. However, a majority of the participating States was in favour of retaining distress signals within the framework of the Rules to give them the widest possible promulgation. A compromise was made by only incorporating within the Regulations the short sentence of Rule 37 which requires a vessel in distress to use the signals and refers to the list of signals given in Annex IV.

Vessels are now specifically required to use one or more of the specified signals when in distress and requiring assistance.

Part E – Exemptions

RULE 38

Exemptions

Any vessel (or class of vessels) provided that she complies with the requirements of the International Regulations for Preventing Collisions at Sea, 1960, the keel of which is laid or which is at a corresponding stage of construction before the entry into force of these Regulations, may be exempted from compliance therewith as follows:

(a) The installation of lights with ranges prescribed in Rule 22, until four years after the date of entry into force of these Regulations.
(b) The installation of lights with colour specifications as prescribed in Section 7 of Annex I to these Regulations, until four years after the date of entry into force of these Regulations.
(c) The repositioning of lights as a result of conversion from Imperial to metric units and rounding off measurement figures, permanent exemption.
(d) (i) The repositioning of masthead lights on vessels of less than 150 metres in length, resulting from the prescriptions of Section 3(a) of Annex I to these Regulations, permanent exemption.
   (ii) The repositioning of masthead lights on vessels of 150 metres or more in length, resulting from the prescriptions of Section 3(a) of Annex I to these Regulations, until nine years after the date of entry into force of these Regulations.
(e) The repositioning of masthead lights resulting from the prescriptions of Section 2(b) of Annex I to these Regulations, until nine years after the date of entry into force of these Regulations.
(f) The repositioning of sidelights resulting from the prescriptions of Sections 2(g) and 3(b) of Annex I to these Regulations, until nine years after the date of entry into force of these Regulations.
(g) The requirements for sound signal appliances prescribed in Annex III to these Regulations, until nine years after the date of entry into force of these Regulations.
(h) The repositioning of all-round lights resulting from the prescription of Section 9(b) of Annex I to these Regulations, permanent exemption.

COMMENT:

This Rule was necessary to allow sufficient time for the required changes to be made in the positions and characteristics of lights, and in the performance of sound signalling appliances. Paragraph (h) was added with the amendments of 1981.

A Guide to the Collision Avoidance Rules. DOI: 10.1016/B978-0-08-097170-4.00005-2

With the amendment of 1987 to Rule 1(e) governments are able to extend the examples mentioned in paragraphs (d)(ii), (e), (f) and (g) (see page 17).

There is, however, no exemption with respect to the carriage of lights and shapes introduced for the first time in the 1972 Regulations. New lights and shapes which are required to be carried by certain vessels include:

a. The yellow towing light prescribed in Rule 24(a)(iv) for a power-driven vessel engaged in towing another vessel.
b. The all-round flashing yellow light (flashing at an increased rate) prescribed in Rule 23(b) for air-cushion vessels operating in the non-displacement mode.
c. The lights required by Rule 27(c) for a vessel engaged in a difficult towing operation.
d. The lights and shapes prescribed in Rule 27(d) for vessels engaged in dredging or underwater operations when an obstruction exists.
e. The rigid replica of the International Code flag 'A' prescribed in Rule 27(e) for a small vessel engaged in diving operations.

Annexes to the Rules

INTRODUCTORY COMMENT:

Annex I

Technical details of lights and shapes, and information about their required positions, are given in Annex I of the Regulations. In 1981 amendments were approved by IMO to the following sections of Annex I: 1; 2(e), (f), (i), (j) and (k); 3(b) and (c); 5; 8 (Note); 9(a) and (b); 10(a) and (b); 13.

Section 2(e) was amended because of the amendment to Rule 24(a)(i) and (c)(i). If the additional masthead light(s) to indicate towing are carried on the aftermast the lowest after masthead light must be carried at least 4.5 metres higher than the forward masthead light.

Section 2(f) was amended so that when three all-round lights are carried in a vertical line, by a vessel restricted in her ability to manœuvre or constrained by her draught, and it is not practicable to carry them below the masthead lights they may be carried above the after masthead light(s) or vertically in between the forward masthead light(s) and the after masthead light(s).

If the three all-round lights are carried vertically in between the forward masthead light(s) and the after masthead light(s) the amendment to Section 3(c) of Annex I requires the all-round light to be placed at a distance of not less than 2 metres from the fore and aft centreline measured in the athwartships horizontal direction.

The other amendments to Annex I, made in 1981, are relatively minor changes most of which were introduced for purposes of clarification.

In 1987 the fifteenth Assembly of IMO adopted amendments to Annex I Sections 2(d), 2(i)(ii), and 10. These were minor amendments to achieve more consistency between the Rules and the text of Annex I.

In 1993 the eighteenth Assembly of IMO adopted amendments to Annex I Sections 3 and 9 to overcome problems which had been experienced in the horizontal positioning of the masthead light on small ships (Section 3) and in the positioning of all-round lights (Section 9). The eighteenth Assembly also adopted an amendment for the addition of a new Section 13 *High speed craft,* which gives provisions for the vertical positioning of the masthead light on high speed craft. The previous Section 13 *Approval* is now Section 14. In 2001 the 22nd Assembly of IMO adopted amendments to Annex I, Section 13 *High speed craft,* to overcome problems which had been experienced in the positioning of masthead lights on certain types of high speed craft.

The Maritime Safety Committee of IMO adopted at its 84th session in 2008 MSC.1/Circ. 1260 on Unified Interpretations of COLREG 72, concerning Annex 1, as follows:

A Guide to the Collision Avoidance Rules. DOI: 10.1016/B978-0-08-097170-4.00006-4

Section 3(b) – Horizontal positioning of lights

The term "near the side" is interpreted as being a distance of not more than 10% of the breadth of the vessel in board from the side, a maximum of 1 metre. Where the application of above requirement is impractical, such as small ships with superstructure of reduced width, exemption may be given on the basis of the Flag Authority acceptance.

Section 9(b) – Horizontal sectors

In order to comply with the 1 mile requirement in 9(b)(ii), the all-round lights shall be screened less than 180 degrees. These lights are not a fixed point but have a certain extension, it may be accepted that all-round lights are screened up to 180 degrees. Screening details are to be considered by Societies when carrying out the drawing process.

Annex II

Additional signals which may be exhibited by fishing vessels fishing in close proximity are listed in Annex II. In 1993 the eighteenth Assembly of IMO adopted amendments to Section 2 *Signals for trawlers* to require vessels of 20 metres or more when engaged in trawling, including pair trawling, to exhibit the prescribed signals when fishing in close proximity. The signals were previously optional for such vessels.

Annex III

Technical details of sound signal appliances are given in Annex III. In 1981 minor amendments were made to the text of Sections 1(d), 2(a), 2(b) and 3 for purposes of clarification. In 2001 amendments were made to the text of Sections 1(a), 1(c) and 2(b), consequential to the amendment of Rule 33(a).

Annex IV

The signals to be used by a vessel which is in distress and requires assistance are listed in Annex IV. This Annex was not amended in 1981.

In 1987 the fifteenth Assembly of IMO adopted an amendment to Annex IV consisting of an additional paragraph (o) to section I. This amendment, covering distress signals transmitted by radiocommunication systems, was deemed to be necessary as a result of the introduction of the Global Maritime Distress and Safety System.

In 1993 the eighteenth Assembly of IMO adopted an amendment to paragraph (o) of Annex IV to include signals from survival craft radar transponders as approved signals transmitted by radiocommunication systems.

In 2007, the 25th Assembly of IMO adopted an amendment to Annex IV *Distress Signals* (A25/Res.1004/Rev.1) to align the distress signals with the requirements of the Global Maritime Distress and Safety System, (see Annex IV, page 148).

Annex I

Positioning and technical details of lights and shapes

1. Definition

The term 'height above the hull' means height above the uppermost continuous deck. This height shall be measured from the position vertically beneath the location of the light.

2. Vertical positioning and spacing of lights

(a) On a power-driven vessel of 20 metres or more in length the masthead lights shall be placed as follows:
  (i) the forward masthead light, or if only one masthead light is carried, then that light, at a height above the hull of not less than 6 metres, and, if the breadth of the vessel exceeds 6 metres, then at a height above the hull not less than such breadth, so however that the light need not be placed at a greater height above the hull than 12 metres;
  (ii) when two masthead lights are carried the after one shall be at least 4.5 metres vertically higher than the forward one.
(b) The vertical separation of masthead lights of power-driven vessels shall be such that in all normal conditions of trim the after light will be seen over and separate from the forward light at a distance of 1,000 metres from the stem when viewed from sea level.
(c) The masthead light of a power-driven vessel of 12 metres but less than 20 metres in length shall be placed at a height above the gunwale of not less than 2.5 metres.
(d) A power-driven vessel of less than 12 metres in length may carry the uppermost light at a height of less than 2.5 metres above the gunwale. When however a masthead light is carried in addition to sidelights and a sternlight or the all-round light prescribed in Rule 23(c)(i) is carried in addition to sidelights, then such masthead light shall be carried at least 1 metre higher than the sidelights.
(e) One of the two or three masthead lights prescribed for a power-driven vessel when engaged in towing or pushing another vessel shall be placed in the same position as either the forward masthead light or the after masthead light; provided that, if carried on the aftermast, the lowest after masthead light shall be at least 4.5 metres vertically higher than the forward masthead light.
(f) (i) The masthead light or lights prescribed in Rule 23(a) shall be so placed as to be above and clear of all other lights and obstructions except as described in sub-paragraph (ii).
  (ii) When it is impracticable to carry the all-round lights prescribed by Rule 27(b)(i) or Rule 28 below the masthead lights, they may be carried above the after masthead light(s) or vertically in between the forward masthead light(s) and after masthead light(s), provided that in the latter case the requirement of Section 3(c) of this Annex shall be complied with.

(g) The sidelights of a power-driven vessel shall be placed at a height above the hull not greater than three quarters of that of the forward masthead light. They shall not be so low as to be interfered with by deck lights.

(h) The sidelights, if in a combined lantern and carried on a power-driven vessel of less than 20 metres in length, shall be placed not less than 1 metre below the masthead light.

(i) When the Rules prescribe two or three lights to be carried in a vertical line, they shall be spaced as follows:

   (i)   on a vessel of 20 metres in length or more such lights shall be spaced not less than 2 metres apart, and the lowest of these lights shall, except where a towing light is required, be placed at a height of not less than 4 metres above the hull;

   (ii)   on a vessel of less than 20 metres in length such lights shall be spaced not less than 1 metre apart and the lowest of these lights shall, except where a towing light is required, be placed at a height of not less than 2 metres above the gunwale;

   (iii) when three lights are carried they shall be equally spaced.

(j) The lower of the two all-round lights prescribed for a vessel when engaged in fishing shall be at a height above the sidelights not less than twice the distance between the two vertical lights.

(k) The forward anchor light prescribed in Rule 30(a)(i), when two are carried, shall not be less than 4.5 metres above the after one. On a vessel of 50 metres or more in length this forward anchor light shall be placed at a height of not less than 6 metres above the hull.

3. Horizontal positioning and spacing of lights[1]

(a) When two masthead lights are prescribed for a power-driven vessel, the horizontal distance between them shall not be less than one half of the length of the vessel but need not be more than 100 metres. The forward light shall be placed not more than one quarter of the length of the vessel from the stem.

(b) On a power-driven vessel of 20 metres or more in length the sidelights shall not be placed in front of the forward masthead lights. They shall be placed at or near the side of the vessel.

(c) When the lights prescribed in Rule 27(b)(i) or Rule 28 are placed vertically between the forward masthead light(s) and the after masthead light(s) these all-round lights shall be placed at a horizontal distance of not less than 2 metres from the fore and aft centreline of the vessel in the athwartship direction.

(d) When only one masthead light is prescribed for a power-driven vessel, this light shall be exhibited forward of amidships; except that a vessel of less than 20 metres in length need not exhibit this light forward of amidships but shall exhibit it as far forward as is practicable.

---

[1] See page 138 regarding unified interpretation on Annex I section 3(b) Horizontal positioning and spacing of lights.

4. *Details of location of direction-indicating lights for fishing vessels, dredgers and vessels engaged in underwater operations*

(a) The light indicating the direction of the outlying gear from a vessel engaged in fishing as prescribed in Rule 26(c)(ii) shall be placed at a horizontal distance of not less than 2 metres and not more than 6 metres away from the two all-round red and white lights. This light shall be placed not higher than the all-round white light prescribed in Rule 26(c)(i) and not lower than the sidelights.

(b) The lights and shapes on a vessel engaged in dredging or underwater operations to indicate the obstructed side and/or the side on which it is safe to pass, as prescribed in Rule 27(d)(i) and (ii), shall be placed at the maximum practical horizontal distance, but in no case less than 2 metres, from the lights or shapes prescribed in Rule 27(b)(i) and (ii). In no case shall the upper of these lights or shapes be at a greater height than the lower of the three lights or shapes prescribed in Rule 27(b)(i) and (ii).

5. Screens for sidelights

The sidelights of vessels of 20 metres or more in length shall be fitted with inboard screens painted matt black, and meeting the requirements of Section 9 of this Annex. On vessels of less than 20 metres in length the sidelights, if necessary to meet the requirements of Section 9 of this Annex, shall be fitted with inboard matt black screens. With a combined lantern, using a single vertical filament and a very narrow division between the green and red sections, external screens need not be fitted.

6. Shapes

(a) Shapes shall be black and of the following sizes:
  (i)   a ball shall have a diameter of not less than 0.6 metre;
  (ii) a cone shall have a base diameter of not less than 0.6 metre and a height equal to its diameter;
  (iii) a cylinder shall have a diameter of at least 0.6 metre and a height of twice its diameter;
  (iv) a diamond shape shall consist of two cones as defined in (ii) above having a common base.

(b) The vertical distance between shapes shall be at least 1.5 metres.

(c) In a vessel of less than 20 metres in length shapes of lesser dimensions but commensurate with the size of the vessel may be used and the distance apart may be correspondingly reduced.

7. Colour specification of lights

The chromaticity of all navigation lights shall conform to the following standards, which lie within the boundaries of the area of the diagram specified for each colour by the International Commission on Illumination (CIE).

The boundaries of the area for each colour are given by indicating the corner co-ordinates, which are as follows:

(i) White						
x	0.525	0.525	0.452	0.310	0.310	0.443
y	0.382	0.440	0.440	0.348	0.283	0.382
(ii) Green						
x	0.028	0.009	0.300	0.203		
y	0.385	0.723	0.511	0.356		
(iii) Red						
x	0.680	0.660	0.735	0.721		
y	0.320	0.320	0.265	0.259		
(iv) Yellow						
x	0.612	0.618	0.575	0.575		
y	0.382	0.382	0.425	0.406		

8. Intensity of lights

(a) The minimum luminous intensity of lights shall be calculated by using the formula:

$$I = 3.43 \times 10^6 \times T \times D^2 \times K^{-D}$$

where I   is luminous intensity in candelas under service conditions,
  T   is threshold factor 2×10^{-7} lux,
  D   is range of visibility (luminous range) of the light in nautical miles,
  K   is atmospheric transmissivity.

For prescribed lights the value of K shall be 0.8, corresponding to a meteorological visibility of approximately 13 nautical miles.

(b) A selection of figures derived from the formula is given in the following table:

Range of visibility (luminous range) of light in nautical miles	Luminous intensity of light in candelas for K = 0.8
1	0.9
2	4.3
3	12
4	27
5	52
6	94

Note: The maximum luminous intensity of navigation lights should be limited to avoid undue glare. This shall not be achieved by a variable control of the luminous intensity.

9 Horizontal sectors[2]

(a) (i) In the forward direction, sidelights as fitted on the vessel shall show the minimum required intensities. The intensities must decrease to reach practical cut-off between 1 degree and 3 degrees outside the prescribed sectors.

   (ii) For sternlights and masthead lights and at 22.5 degrees abaft the beam for sidelights, the minimum required intensities shall be maintained over the arc of the horizon up to 5 degrees within the limits of the sectors prescribed in Rule 21. From 5 degrees within the prescribed sectors the intensity may decrease by 50 per cent up to the prescribed limits; it shall decrease steadily to reach practical cut-off at not more than 5 degrees outside the prescribed sectors.

(b) (i) All-round lights shall be so located as not to be obscured by masts, topmasts or structures within angular sectors of more than 6 degrees, except anchor lights prescribed in Rule 30, which need not be placed at an impracticable height above the hull.

   (ii) If it is impracticable to comply with paragraph (b)(i) of this section by exhibiting only one all-round light, two all-round lights shall be used suitably positioned or screened so that they appear, as far as practicable, as one light at a distance of one mile.

10. Vertical sectors

(a) The vertical sectors of electric lights as fitted, with the exception of lights on sailing vessels underway shall ensure that:

   (i) at least the required minimum intensity is maintained at all angles from 5 degrees above to 5 degrees below the horizontal;

   (ii) at least 60 per cent of the required minimum intensity is maintained from 7.5 degrees above to 7.5 degrees below the horizontal.

(b) In the case of sailing vessels underway the vertical sectors of electric lights as fitted shall ensure that:

   (i) at least the required minimum intensity is maintained at all angles from 5 degrees above to 5 degrees below the horizontal;

   (ii) at least 50 per cent of the required minimum intensity is maintained from 25 degrees above to 25 degrees below the horizontal.

(c) In the case of lights other than electric these specifications shall be met as closely as possible.

11. Intensity of non-electric lights

Non-electric lights shall so far as practicable comply with the minimum intensities, as specified in the Table given in Section 8 of this Annex.

---

[2] See page 138 regarding unified interpretation on Annex I section 9(b) Horizontal sectors.

12. Manœuvring light

Notwithstanding the provisions of paragraph 2(f) of this Annex the manœuvring light described in Rule 34(b) shall be placed in the same fore and aft vertical plane as the masthead light or lights and, where practicable, at a minimum height of 2 metres vertically above the forward masthead light, provided that it shall be carried not less than 2 metres vertically above or below the after masthead light. On a vessel where only one masthead light is carried the manœuvring light, if fitted, shall be carried where it can best be seen, not less than 2 metres vertically apart from the masthead light.

13. High speed craft

(a) The masthead light of high-speed craft may be placed at a height related to the breadth of the craft lower than that prescribed in paragraph 2(a)(i) of this Annex, provided that the base angle of the isosceles triangle formed by the sidelights and masthead light, when seen in end elevation, is not less than 27°.

(b) On high-speed craft of 50 metres or more in length, the vertical separation between fore mast and main mast light of 4.5 metres required by paragraph 2(a)(ii) of this Annex may be modified provided that such distance shall not be less than the value determined by the following formula:

$$y = \frac{(a + 17\Psi)C}{1000} + 2$$

where: y   is the height of the mainmast light above the foremast light in metres;
      a   is the height of the foremast light above the water surface in service condition in metres;
      Ψ is the trim in service condition in degrees;
      C   is the horizontal separation of masthead lights in metres.

14. Approval

The construction of lights and shapes and the installation of lights on board the vessel shall be to the satisfaction of the appropriate authority of the State whose flag the vessel is entitled to fly.

Annex II

Additional signals for fishing vessels fishing in close proximity

1. General

The lights mentioned herein shall, if exhibited in pursuance of Rule 26(d), be placed where they can best be seen. They shall be at least 0.9 metre apart but at a lower level than lights prescribed in Rule 26(b)(i) and (c)(i). The lights shall be visible all round the horizon at a distance of at least 1 mile but at a lesser distance than the lights prescribed by these Rules for fishing vessels.

2. Signals for trawlers

(a) Vessels of 20 metres or more in length when engaged in trawling, whether using demersal or pelagic gear, shall exhibit:
   (i)   when shooting their nets:
         two white lights in a vertical line;
   (ii) when hauling their nets:
         one white light over one red light in a vertical line;
   (iii) when the net has come fast upon an obstruction:
         two red lights in a vertical line.
(b) Each vessel of 20 metres or more in length engaged in pair trawling shall exhibit:
   (i)   by night, a searchlight directed forward and in the direction of the other vessel of the pair;
   (ii) when shooting or hauling their nets or when their nets have come fast upon an obstruction, the lights prescribed in 2(a) above.
(c) A vessel of less than 20 metres in length engaged in trawling, whether using demersal or pelagic gear or engaged in pair trawling, may exhibit the lights prescribed in paragraphs (a) or (b) of this section, as appropriate.

3. Signals for purse seiners

Vessels engaged in fishing with purse seine gear may exhibit two yellow lights in a vertical line. These lights shall flash alternately every second and with equal light and occultation duration. These lights may be exhibited only when the vessel is hampered by its fishing gear.

Annex III

Technical details of sound signal appliances

1. Whistles

(a) *Frequencies and range of audibility*
   The fundamental frequency of the signal shall lie within the range 70–700 Hz. The range of audibility of the signal from a whistle shall be determined by those frequencies, which may include the fundamental and/or one or more higher frequencies, which lie within the range 180–700 Hz (± 1 per cent) for a vessel of 20 metres or more in length, or 180–2100 Hz (± 1 per cent) for a vessel of less than 20 metres in length and which provide the sound pressure levels specified in paragraph 1(c) below.

(b) *Limits of fundamental frequencies*
   To ensure a wide variety of whistle characteristics, the fundamental frequency of a whistle shall be between the following limits:
   (i)   70–200 Hz, for a vessel 200 metres or more in length;
   (ii) 130–350 Hz, for a vessel 75 metres but less than 200 metres in length;
   (iii) 250–700 Hz, for a vessel less than 75 metres in length.

(c) *Sound signal intensity and range of audibility*
   A whistle fitted in a vessel shall be provided, in the direction of maximum intensity of the whistle and at a distance of 1 metre from it, a sound pressure level in at least one 1/3rd-octave band within the range of frequencies 180–700 Hz (± 1 per cent) for a vessel of 20 metres or more in length, or 180–2100 Hz (± 1 per cent) for a vessel of less than 20 metres in length, of not less than the appropriate figure given in the table below.
   The range of audibility in the table below is for information and is approximately the range at which a whistle may be heard on its forward axis with 90 per cent probability in conditions of still air on board a vessel having average background noise level at the listening posts (taken to be 68 dB in the octave band centred on 250 Hz and 63 dB in the octave band centred on 500 Hz).
   In practice the range at which a whistle may be heard is extremely variable and depends critically on weather conditions; the values given can be regarded as typical but under conditions of strong wind or high ambient noise level at the listening post the range may be much reduced.

Length of vessel in metres	1/3-octave band level at 1 metre in dB referred to 2 X 10^{-5} N/m^2	Audibility range in nautical miles
200 or more	143	2
75 but less than 200	138	1.5
20 but less than 75	130	1
Less than 20	120[1]	0.5
	115[2]	
	111[3]	

[1] When the measured frequencies lie within the range 180–450 Hz
[2] When the measured frequencies lie within the range 450–800 Hz
[3] When the measured frequencies lie within the range 800–2100 Hz

(d) *Directional properties*

The sound pressure level of a directional whistle shall be not more than 4 dB below the prescribed sound pressure level on the axis at any direction in the horizontal plane within ± 45 degrees of the axis. The sound pressure level at any other direction in the horizontal plane shall be not more than 10 dB below the prescribed sound pressure level on the axis, so that the range in any direction will be at least half the range on the forward axis. The second pressure level shall be measured in that 1/3rd-octave band which determines the audibility range.

(e) *Positioning of whistles*

When a directional whistle is to be used as the only whistle on a vessel, it shall be installed with its maximum intensity directed straight ahead.

A whistle shall be placed as high as practicable on a vessel, in order to reduce interception of the emitted sound by obstructions and also to minimise hearing damage risk to personnel.

The sound pressure level of the vessel's own signal at listening posts shall not exceed 110 dB (A) and so far as practicable should not exceed 100 dB (A).

(f) *Fitting of more than one whistle*

If whistles are fitted at a distance apart of more than 100 metres, it shall be so arranged that they are not sounded simultaneously.

(g) *Combined whistle systems*

If due to the presence of obstructions the sound field of a single whistle or of one of the whistles referred to in paragraph 1(f) above is likely to have a zone of greatly reduced signal level, it is recommended that a combined whistle system be fitted so as to overcome this reduction. For the purposes of the Rules a combined whistle system is to be regarded as a single whistle. The whistles of a combined system shall be located at a distance apart of not more than 100 metres and arranged to be sounded simultaneously. The frequency of any one whistle shall differ from those of the others by at least 10 Hz.

2. Bell or gong

(a) *Intensity of signal*

A bell or gong, or other device having similar sound characteristics shall produce a sound pressure level of not less than 110 dB at a distance of 1 metre from it.

(b) *Construction*

Bells and gongs shall be made of corrosion-resistant material and designed to give a clear tone. The diameter of the mouth of the bell shall be not less than 300 mm for vessels of 20 metres or more in length. Where practicable, a power-driven bell striker is recommended to ensure constant force but manual operation shall be possible. The mass of the striker shall be not less than 3 per cent of the mass of the bell.

3. Approval

The construction of sound signal appliances, their performance and their installation on board the vessel shall be to the satisfaction of the appropriate authority of the State whose flag the vessel is entitled to fly.

Annex IV

Distress signals

1 The following signals, used or exhibited either together or separately, indicate distress and need of assistance:

(a) a gun or other explosive signal fired at intervals of about a minute;

(b) a continuous sounding with any fog-signalling apparatus;

(c) rockets or shells, throwing red stars fired one at a time at short intervals;

(d) a signal made by radiotelegraphy or by any other signalling method consisting of the group ... _ _ _ ... (SOS) in the Morse Code;

(e) a signal sent by radiotelephony consisting of the spoken word 'Mayday';

(f) the International Code Signal of distress indicated by NC;

(g) a signal consisting of a square flag having above or below it a ball or anything resembling a ball;

(h) flames on the vessel (as from a burning tar barrel, oil barrel, etc.);

(i) a rocket parachute flare or a hand flare showing a red light;

(j) a smoke signal giving off orange-coloured smoke;

(k) slowly and repeatedly raising and lowering arms outstretched to each side;

(l) a distress alert by means of digital selective calling (DSC) transmitted on:
   (a) VHF channel 70, or
   (b) MF/HF on the frequencies 2187.5 kHz, 8414.5 kHz, 4207.5 kHz, 6312 kHz, 12577 kHz or 16804.5 kHz;

(m) a ship-to-shore distress alert transmitted by the ship's INMARSAT or other mobile satellite service provider ship earth station;

(n) signals transmitted by emergency position-indicating radio beacons;

(o) approved signals transmitted by radiocommunication systems, including survival craft radar transponders.

2 The use or exhibition of any of the foregoing signals except for the purpose of indicating distress and need of assistance and the use of other signals which may be confused with any of the above signals is prohibited.

3 Attention is drawn to the relevant sections of the International Code of Signals, the International Aeronautical and Maritime Search and Rescue Manual, Volume III and the following signals:

(a) a piece of orange-coloured canvas with either a black square and circle or other appropriate symbol (for identification from the air);

(b) a dye marker.

International convention on standards of training, certification and watchkeeping for seafarers, 1978

Regulation II/1

Basic principles to be observed in keeping a navigational watch

1. Parties shall direct the attention of shipowners, ship operators, masters and watchkeeping personnel to the following principles which shall be observed to ensure that a safe navigational watch is maintained at all times.
2. The master of every ship is bound to ensure that watchkeeping arrangements are adequate for maintaining a safe navigational watch. Under the master's general direction, the officers of the watch are responsible for navigating the ship safely during their periods of duty when they will be particularly concerned with avoiding collision and stranding.
3. The basic principles, including but not limited to the following, shall be taken into account on all ships.
4. *Watch arrangements*
   (a) The composition of the watch shall at all times be adequate and appropriate to the prevailing circumstances and conditions and shall take into account the need for maintaining a proper look-out.
   (b) When deciding the composition of the watch on the bridge which may include appropriate deck ratings, the following factors, *inter alia,* shall be taken into account:
     (i) at no time shall the bridge be left unattended;
     (ii) weather conditions, visibility and whether there is daylight or darkness;
     (iii) proximity of navigational hazards which may make it necessary for the officer in charge of the watch to carry out additional navigational duties;
     (iv) use and operational condition of navigational aids such as radar or electronic position-indicating devices and any other equipment affecting the safe navigation of the ship;
     (v) whether the ship is fitted with automatic steering;
     (vi) any unusual demands on the navigational watch that may arise as a result of special operational circumstances.
5. *Fitness for duty*
   The watch system shall be such that the efficiency of watchkeeping officers and watchkeeping ratings is not impaired by fatigue. Duties shall be so organized that the first watch at the commencement of a voyage and the subsequent relieving watches are sufficiently rested and otherwise fit for duty.

A Guide to the Collision Avoidance Rules. DOI: 10.1016/B978-0-08-097170-4.00007-6

6. *Navigation*
   (a) The intended voyage shall be planned in advance taking into consideration all pertinent information and any course laid down shall be checked before the voyage commences.
   (b) During the watch the course steered, position and speed shall be checked at sufficiently frequent intervals, using any available navigational aids necessary, to ensure that the ship follows the planned course.
   (c) The officer of the watch shall have full knowledge of the location and operation of all safety and navigational equipment on board the ship and shall be aware and take account of the operating limitations of such equipment.
   (d) The officer in charge of a navigational watch shall not be assigned or undertake any duties which would interfere with the safe navigation of the ship.

7. *Navigational equipment*
   (a) The officer of the watch shall make the most effective use of all navigational equipment at his disposal.
   (b) When using radar, the officer of the watch shall bear in mind the necessity to comply at all times with the provisions on the use of radar contained in the applicable regulations for preventing collisions at sea.
   (c) In cases of need the officer of the watch shall not hesitate to use the helm, engines and sound signalling apparatus.

8. *Navigational duties and responsibilities*
   (a) The officer in charge of the watch shall:
       (i)   keep his watch on the bridge which he shall in no circumstances leave until properly relieved;
       (ii) continue to be responsible for the safe navigation of the ship, despite the presence of the master on the bridge, until the master informs him specifically that he has assumed that responsibility and this is mutually understood;
       (iii) notify the master when in any doubt as to what action to take in the interest of safety;
       (iv) not hand over the watch to the relieving officer if he has reason to believe that the latter is obviously not capable of carrying out his duties effectively, in which case he shall notify the master accordingly.
   (b) On taking over the watch the relieving officer shall satisfy himself as to the ship's estimated or true position and confirm its intended track, course and speed and shall note any dangers to navigation expected to be encountered during his watch.
   (c) A proper record shall be kept of the movements and activities during the watch relating to the navigation of the ship.

9. *Look-out*
   In addition to maintaining a proper look-out for the purpose of fully appraising the situation and the risk of collision, stranding and other dangers to navigation, the duties of the look-out shall include the detection of ships or aircraft in

distress, shipwrecked persons, wrecks and debris. In maintaining a look-out the following shall be observed:

(a) the look-out must be able to give full attention to the keeping of a proper look-out and no other duties shall be undertaken or assigned which could interfere with that task;

(b) the duties of the look-out and helmsman are separate and the helmsman shall not be considered to be the look-out while steering, except in small ships where an unobstructed all-round view is provided at the steering position and there is no impairment of night vision or other impediment to the keeping of a proper look-out. The officer in charge of the watch may be the sole look-out in daylight provided that on each such occasion:

  (i) the situation has been carefully assessed and it has been established without doubt that it is safe to do so;

  (ii) full account has been taken of all relevant factors including, but not limited to:
  - state of weather
  - visibility
  - traffic density
  - proximity of danger to navigation
  - the attention necessary when navigating in or near traffic separation schemes;

  (iii) assistance is immediately available to be summoned to the bridge when any change in the situation so requires.

10. *Navigation with pilot embarked*

Despite the duties and obligations of a pilot, his presence on board does not relieve the master or officer in charge of the watch from their duties and obligations for the safety of the ship. The master and the pilot shall exchange information regarding navigation procedures, local conditions and the ship's characteristics. The master and officer of the watch shall co-operate closely with the pilot and maintain an accurate check of the ship's position and movement.

11. *Protection of the marine environment*

The master and officer in charge of the watch shall be aware of the serious effects of operational or accidental pollution of the marine environment and shall take all possible precautions to prevent such pollution, particularly within the framework of relevant international and port regulations.

IMO Recommendation on navigational watchkeeping

Recommendation on operational guidance for officers in charge of a navigational watch

Introduction

1. This Recommendation contains operational guidance of general application for officers in charge of a navigational watch, which masters are expected to supplement as appropriate. It is essential that officers of the watch appreciate that the efficient performance of their duties is necessary in the interests of the safety of life and property at sea and the prevention of pollution of the marine environment.

General

2. The officer of the watch is the master's representative and his primary responsibility at all times is the safe navigation of the ship. He should at all times comply with the applicable regulations for preventing collisions at sea (see also paragraphs 22 and 23).
3. It is of special importance that at all times the officer of the watch ensures that an efficient look-out is maintained. In a ship with a separate chart room the officer of the watch may visit the chart room, when essential, for a short period for the necessary performance of his navigational duties, but he should previously satisfy himself that it is safe to do so and ensure that an efficient look-out is maintained.
4. The officer of the watch should bear in mind that the engines are at his disposal and he should not hesitate to use them in case of need. However, timely notice of intended variations of engine speed should be given where possible. He should also know the handling characteristics of his ship, including its stopping distance, and should appreciate that other ships may have different handling characteristics.
5. The officer of the watch should also bear in mind that the sound signalling apparatus is at his disposal and he should not hesitate to use it in accordance with the applicable regulations for preventing collisions at sea.

Taking over the navigational watch

6. The relieving officer of the watch should ensure that members of his watch are fully capable of performing their duties, particularly as regards their adjustment to night vision.

A Guide to the Collision Avoidance Rules. DOI: 10.1016/B978-0-08-097170-4.00008-8

7. The relieving officer should not take over the watch until his vision is fully adjusted to the light conditions and he has personally satisfied himself regarding:

(a) standing orders and other special instructions of the master relating to navigation of the ship;

(b) position, course, speed and draught of the ship;

(c) prevailing and predicted tides, currents, weather, visibility and the effect of these factors upon course and speed;

(d) navigational situation, including but not limited to the following:

   (i) operational condition of all navigational and safety equipment being used or likely to be used during the watch;

   (ii) errors of gyro and magnetic compasses;

   (iii) presence and movement of ships in sight or known to be in the vicinity;

   (iv) conditions and hazards likely to be encountered during his watch;

   (v) possible effects of heel, trim, water density and squat* on underkeel clearance.

8. If at any time the officer of the watch is to be relieved a manœuvre or other action to avoid any hazard is taking place, the relief of the officer should be deferred until such action has been completed.

Periodic checks of navigational equipment

9. Operational tests of shipboard navigational equipment should be carried out at sea as frequently as practicable and as circumstances permit, in particular when hazardous conditions affecting navigation are expected; where appropriate these tests should be recorded.

10. The officer of the watch should make regular checks to ensure that:

(a) the helmsman or the automatic pilot is steering the correct course;

(b) the standard compass error is determined at least once a watch and, when possible, after any major alteration of course; the standard and gyro-compasses are frequently compared and repeaters are synchronized with their master compass;

(c) the automatic pilot is tested manually at least once a watch;

(d) the navigation and signal lights and other navigational equipment are functioning properly.

Automatic pilot

11. The officer of the watch should bear in mind the necessity to comply at all times with the requirements of Regulation 19, Chapter V of the International Convention for the Safety of Life at Sea, 1974. He should take into account the need to station the helmsman and to put the steering into manual control in good time to allow any potentially hazardous situation to be dealt with in a safe

---

* Squat: The decrease in clearance beneath the ship which occurs when the ship moves through the water and is caused both by bodily sinkage and by change of trim. The effect is accentuated in shallow water and is reduced with a reduction in ship's speed.

manner. With a ship under automatic steering it is highly dangerous to allow a situation to develop to the point where the officer of the watch is without assistance and has to break the continuity of the look-out in order to take emergency action. The change-over from automatic to manual steering and vice-versa should be made by, or under the supervision of, a responsible officer.

Electronic navigational aids

12. The officer of the watch should be thoroughly familiar with the use of electronic navigational aids carried, including their capabilities and limitations.
13. The echo-sounder is a valuable navigational aid and should be used whenever appropriate.

Radar

14. The officer of the watch should use the radar when appropriate and whenever restricted visibility is encountered or expected, and at all times in congested waters having due regard to its limitations.
15. Whenever radar is in use, the officer of the watch should select an appropriate range scale, observe the display carefully and plot effectively.
16. The officer of the watch should ensure that range scales employed are changed at sufficiently frequent intervals so that echoes are detected as early as possible.
17. It should be borne in mind that small or poor echoes may escape detection.
18. The officer of the watch should ensure that plotting or systematic analysis is commenced in ample time.
19. In clear weather, whenever possible, the officer of the watch should carry out radar practice.

Navigation in coastal waters

20. The largest scale chart on board, suitable for the area and corrected with the latest available information, should be used. Fixes should be taken at frequent intervals; whenever circumstances allow, fixing should be carried out by more than one method.
21. The officer of the watch should positively identify all relevant navigation marks.

Clear weather

22. The officer of the watch should take frequent and accurate compass bearings of approaching ships as a means of early detection of risk of collision; such risk may sometimes exist even when an appreciable bearing change is evident, particularly when approaching a very large ship or a tow or when approaching a ship at close range. He should also take early and positive action in compliance with the

applicable regulations for preventing collisions at sea and subsequently check that such action is having the desired effect.

Restricted visibility

23. When restricted visibility is encountered or expected, the first responsibility of the officer of the watch is to comply with the relevant rules of the applicable regulations for preventing collisions at sea, with particular regard to the sounding of fog signals, proceeding at a safe speed and having the engines ready for immediate manœuvres. In addition, he should:

(a) inform the master (see paragraph 24);

(b) post a proper look-out and helmsman and, in congested waters, revert to hand steering immediately;

(c) exhibit navigation lights;

(d) operate and use the radar.

It is important that the officer of the watch should know the handling characteristics of his ship, including its stopping distance, and should appreciate that other ships may have different handling characteristics.

Calling the master

24. The officer of the watch should notify the master immediately in the following circumstances;

(a) if restricted visibility is encountered or expected;

(b) if the traffic conditions or the movements of other ships are causing concern;

(c) if difficulty is experienced in maintaining course;

(d) on failure to sight land, a navigation mark or to obtain soundings by the expected time;

(e) if, unexpectedly, land or a navigation mark is sighted or change in sounding occurs;

(f) on the breakdown of the engines, steering gear or any essential navigational equipment;

(g) in heavy weather if in any doubt about the possibility of weather damage;

(h) if the ship meets any hazard to navigation, such as ice or derelicts;

(i) in any other emergency or situation in which he is in any doubt.

Despite the requirement to notify the master immediately in the foregoing circumstances, the officer of the watch should in addition not hesitate to take immediate action for the safety of the ship, where circumstances so require.

Navigation with pilot embarked

25. If the officer of the watch is in any doubt as to the pilot's actions or intentions, he should seek clarification from the pilot; if doubt still exists, he should notify the master immediately and take whatever action is necessary before the master arrives.

Watchkeeping personnel

26. The officer of the watch should give watchkeeping personnel all appropriate instructions and information which will ensure the keeping of a safe watch including an appropriate look-out.

Ship at anchor

27. If the master considers it necessary, a continuous navigational watch should be maintained at anchor. In all circumstances, while at anchor, the officer of the watch should:

(a) determine and plot the ship's position on the appropriate chart as soon as practicable; when circumstances permit, check at sufficiently frequent intervals whether the ship is remaining securely at anchor by taking bearings of fixed navigational marks or readily identifiable shore objects;

(b) ensure that an efficient look-out is maintained;

(c) ensure that inspection rounds of the ship are made periodically;

(d) observe meteorological and tidal conditions and the state of the sea;

(e) notify the master and undertake all necessary measures if the ship drags anchor;

(f) ensure that the state of readiness of the main engines and other machinery is in accordance with the master's instructions;

(g) if visibility deteriorates, notify the master and comply with the applicable regulations for preventing collisions at sea;

(h) ensure that the ship exhibits the appropriate lights and shapes and that appropriate sound signals are made at all times, as required;

(i) take measures to protect the environment from pollution by the ship and comply with applicable pollution regulations.

International convention on standards of training, certification and watchkeeping for seafarers, 1978, as amended

Annex

Chapter VIII/1

Watchkeeping

Regulation VIII/1

Fitness for duty

Each Administration shall, for the purpose of preventing fatigue:

1. establish and enforce rest periods for watchkeeping personnel; and
2. require that watch systems are so arranged that the efficiency of all watchkeeping personnel is not impaired by fatigue and that duties are so organized that the first watch at the commencement of a voyage and subsequent relieving watches are sufficiently rested and otherwise fit for duty.

Regulation VIII/2

Watchkeeping arrangements and principles to be observed

1. Administrations shall direct the attention of companies, masters, chief engineer officers and all watchkeeping personnel to the requirements, principles and guidance set out in the STCW Code which shall be observed to ensure that a safe continuous watch or watches appropriate to the prevailing circumstances and conditions are maintained in all seagoing ships at all times.
2. Administrations shall require the master of every ship to ensure that watchkeeping arrangements are adequate for maintaining a safe watch or watches, taking into account the prevailing circumstances and conditions and that, under the master's general direction:
   (1) officers in charge of the navigational watch are responsible for navigating the ship safely during their periods of duty, when they shall be physically present on the navigating bridge or in a directly associated location such as the chartroom or bridge control room at all times;
   (2) radio operators are responsible for maintaining a continuous radio watch on appropriate frequencies during their periods of duty;

A Guide to the Collision Avoidance Rules. DOI: 10.1016/B978-0-08-097170-4.00009-X

(3) officers in charge of an engineering watch, as defined in the STCW Code and under the direction of the chief engineer officer, shall be immediately available and on call to attend the machinery spaces and, when required, shall be physically present in the machinery space during their periods of responsibility; and

(4) an appropriate and effective watch or watches are maintained for the purpose of safety at all times, while the ship is at anchor or moored and, if the ship is carrying hazardous cargo, the organization of such watch or watches takes full account of the nature, quantity, packing and stowage of the hazardous cargo and of any special conditions prevailing on board, afloat or ashore.

STCW Code

Part A

Mandatory Standards Regarding Provisions of the Annex to the Convention

Chapter VIII

Standards Regarding Watchkeeping

Part 3: Watchkeeping at sea

Part 3.1: Principles to be observed in keeping a navigational watch

12. The officer in charge of the navigational watch is the master's representative and is primarily responsible at all times for the safe navigation of the ship and for complying with the International Regulations for Preventing Collisions at Sea, 1972.

Look-out

13. A proper look-out shall be maintained at all times in compliance with rule 3 of the International Regulations for Preventing Collisions at Sea, 1972 and shall serve the purpose of:
   1. maintaining a continuous state of vigilance by sight and hearing as well as by all other available means, with regard to any significant change in the operating environment;
   2. fully appraising the situation and the risk of collision, stranding and other dangers to navigation; and
   3. detecting ships or aircraft in distress, shipwrecked persons, wrecks, debris and other hazards to safe navigation.
14. The look-out must be able to give full attention to the keeping of a proper look-out and no other duties shall be undertaken or assigned which could interfere with that task.
15. The duties of the look-out and helmsperson are separate and the helmsperson shall not be considered to be the look-out while steering, except in small ships where an unobstructed all-round view is provided at the steering position and there is no impairment of night vision or other impediment to the keeping of a proper look-out. The officer in charge of the navigational watch may be the sole look-out in daylight provided that on each such occasion:
   1. the situation has been carefully assessed and it has been established without doubt that it is safe to do so;
   2. full account has been taken of all relevant factors including, but not limited to:
      • state of weather,
      • visibility,
      • traffic density,

- proximity of dangers to navigation, and
- the attention necessary when navigating in or near traffic separation schemes; and

3. assistance is immediately available to be summoned to the bridge when any change in the situation so requires.

16. In determining that the composition of the navigational watch is adequate to ensure that a proper look-out can continuously be maintained, the master shall take into account all relevant factors, including those described in this section of the Code, as well as the following factors:

   1. visibility, state of weather and sea;
   2. traffic density, and other activities occurring in the area in which the vessel is navigating;
   3. the attention necessary when navigating in or near traffic separation schemes or other routeing measures;
   4. the additional workload caused by the nature of the ship's functions, immediate operating requirements and anticipated manœuvres;
   5. the fitness for duty of any crew members on call who are assigned as members of the watch;
   6. knowledge of, and confidence in, the professional competence of the ship's officers and crew;
   7. the experience of each officer of the watch, and the familiarity of that officer with the ship's equipment, procedures and manœuvring capability;
   8. activities taking place on board the ship at any particular time, including radiocommunication activities and the availability of assistance to be summoned immediately to the bridge when necessary;
   9. the operational status of bridge instrumentation and controls, including alarm systems;
   10. rudder and propeller control and ship manœuvring characteristics;
   11. the size of the ship and the field of vision available from the conning position;
   12. the configuration of the bridge, to the extent that such configuration might inhibit a member of the watch from detecting by sight or hearing any external development; and
   13. any other relevant standard, procedure or guidance relating to watchkeeping arrangements and fitness for duty which has been adopted by the Organization.

Watch arrangements

17. When deciding the composition of the watch on the bridge, which may include appropriately qualified ratings, the following factors, *inter alia,* shall be taken into account:

   1. at no time shall the bridge be left unattended;
   2. weather conditions, visibility and whether there is daylight or darkness;

3. proximity of navigational hazards which may make it necessary for the officer in charge of the watch to carry out additional navigational duties;
4. use and operational condition of navigational aids such as radar or electronic position-indicating devices and any other equipment affecting the safe navigation of the ship;
5. whether the ship is fitted with automatic steering;
6. whether there are radio duties to be performed;
7. unmanned machinery space (UMS) controls, alarms and indicators provided on the bridge, procedures for their use and limitations; and
8. any unusual demands on the navigational watch that may arise as a result of special operational circumstances.

Taking over the watch

18. The officer in charge of the navigational watch shall not hand over the watch to the relieving officer if there is reason to believe that the latter is not capable of carrying out the watchkeeping duties effectively, in which case the master shall be notified.

19. The relieving officer shall ensure that the members of the relieving watch are fully capable of performing their duties, particularly as regards their adjustment to night vision. Relieving officers shall not take over the watch until their vision is fully adjusted to the light conditions.

20. Prior to taking over the watch relieving officers shall satisfy themselves as to the ship's estimated or true position and confirm its intended track, course and speed and UMS controls as appropriate, and shall note any dangers to navigation expected to be encountered during their watch.

21. Relieving officers shall personally satisfy themselves regarding the:
1. standing orders and other special instructions of the master relating to navigation of the ship;
2. position, course, speed and draught of the ship;
3. prevailing and predicted tides, currents, weather, visibility and the effect of these factors upon course and speed;
4. procedures for the use of main engines to manœuvre when the main engines are on bridge control; and
5. navigational situation, including but not limited to:
   5.1. the operational condition of all navigational and safety equipment being used or likely to be used during the watch,
   5.2. the errors of gyro and magnetic compasses,
   5.3. the presence and movement of ships in sight or known to be in the vicinity,
   5.4. the conditions and hazards likely to be encountered during the watch, and
   5.5. the possible effects of heel, trim, water density and squat on under keel clearance.

22. If at any time the officer in charge of the navigational watch is to be relieved when a manœuvre or other action to avoid any hazard is taking place, the relief of that officer shall be deferred until such action has been completed.

Performing the navigational watch

23. The officer in charge of the navigational watch shall:
1. keep the watch on the bridge;
2. in no circumstances leave the bridge until properly relieved;
3. continue to be responsible for the safe navigation of the ship, despite the presence of the master on the bridge, until informed specifically that the master has assumed that responsibility and this is mutually understood; and
4. notify the master when in any doubt as to what action to take in the interest of safety.

24. During the watch the course steered, position and speed shall be checked at sufficiently frequent intervals, using any available navigational aids necessary, to ensure that the ship follows the planned course.

25. The officer in charge of the navigational watch shall have full knowledge of the location and operation of all safety and navigational equipment on board the ship and shall be aware and take account of the operating limitations of such equipment.

26. The officer in charge of the navigational watch shall not be assigned or undertake any duties which would interfere with the safe navigation of the ship.

27. Officers of the navigational watch shall make the most effective use of all navigational equipment at their disposal.

28. When using radar, the officer in charge of the navigational watch shall bear in mind the necessity to comply at all times with the provisions on the use of radar contained in the applicable International Regulations for Preventing Collisions at Sea, 1972.

29. In cases of need the officer in charge of the navigational watch shall not hesitate to use the helm, engines and sound signalling apparatus. However, timely notice of intended variations of engine speed shall be given where possible or effective use made of UMS engine controls provided on the bridge in accordance with the applicable procedures.

30. Officers of the navigational watch shall know the handling characteristics of their ship, including its stopping distances, and should appreciate that other ships may have different handling characteristics.

31. A proper record shall be kept during the watch of the movements and activities relating to the navigation of the ship.

32. It is of special importance that at all times the officer in charge of the watch ensures that a proper look-out is maintained. In a ship with a separate chart room the officer in charge of the watch may visit the chart room, when essential, for a short period for the necessary performance of navigational duties, but shall first ensure that it is safe to do so and that proper look-out is maintained.

33. Operational tests of shipboard navigational equipment shall be carried out at sea as frequently as practicable and as circumstances permit, in particular before hazardous conditions affecting navigation are expected. Whenever appropriate, these tests shall be recorded. Such tests shall also be carried out prior to port arrival and departure.

34. The officer in charge of the navigational watch shall make regular checks to ensure that:

1. the person steering the ship or the automatic pilot is steering the correct course;

2. the standard compass error is determined at least once a watch and, when possible, after any major alteration of course; the standard and gyro-compasses are frequently compared and repeaters are synchronized with their master compass;

3. the automatic pilot is tested manually at least once a watch;

4. the navigation and signal lights and other navigational equipment are functioning properly;

5. the radio equipment is functioning properly in accordance with paragraph 86 of this section, and

6. the UMS controls, alarms and indicators are functioning properly.

35. The officer in charge of the navigational watch shall bear in mind the necessity to comply at all times with the current requirements of the International Convention for the Safety of Life at Sea (SOLAS) 1974. The officer of the watch shall take into account:

1. the need to station a person to steer the ship and to put the steering into manual control in good time to allow any potentially hazardous situation to be dealt with in a safe manner; and

2. that with a ship under automatic steering it is highly dangerous to allow a situation to develop to the point where the officer in charge of the watch is without assistance and has to break the continuity of the look-out in order to take emergency action.

36. Officers of the navigational watch shall be thoroughly familiar with the use of all electronic navigational aids carried, including their capabilities and limitations, and shall use each of these aids when appropriate and shall bear in mind that the echo-sounder is a valuable navigational aid.

37. The officer in charge of the navigational watch shall use the radar whenever restricted visibility is encountered or expected, and at all times in congested waters having due regard to its limitations.

38. The officer in charge of the navigational watch shall ensure that range scales employed are changed at sufficiently frequent intervals so that echoes are detected as early as possible. It shall be borne in mind that small or poor echoes may escape detection.

39. Whenever radar is in use, the officer in charge of the navigational watch shall select an appropriate range scale and observe the display carefully, and shall ensure that plotting or systematic analysis is commenced in ample time.

40. The officer in charge of the navigational watch shall notify the master immediately:

1. if restricted visibility is encountered or expected;

2. if the traffic conditions or the movements of other ships are causing concern;

3. if difficulty is experienced in maintaining course;

   4. on failure to sight land, a navigation mark or to obtain soundings by the expected time;
   5. if, unexpectedly, land or a navigation mark is sighted or a change in soundings occurs;
   6. on breakdown of the engines, propulsion machinery remote control, steering gear or any essential navigational equipment, alarm or indicator;
   7. if the radio equipment malfunctions;
   8. in heavy weather, if in any doubt about the possibility of weather damage;
   9. if the ship meets any hazard to navigation, such as ice or a derelict; and
   10. in any other emergency or if in any doubt.
41. Despite the requirement to notify the master immediately in the foregoing circumstances, the officer in charge of the navigational watch shall in addition not hesitate to take immediate action for the safety of the ship, where circumstances so require.
42. The officer in charge of the navigational watch shall give watch-keeping personnel all appropriate instructions and information which will ensure the keeping of a safe watch, including a proper look-out.

Watchkeeping under different conditions and in different areas

Clear weather

43. The officer in charge of the navigational watch shall take frequent and accurate compass bearings of approaching ships as a means of early detection of risk of collision and bear in mind that such risk may sometimes exist even when an appreciable bearing change is evident, particularly when approaching a very large ship or a tow or when approaching a ship at close range. The officer in charge of the navigational watch shall also take early and positive action in compliance with the applicable regulations for preventing collisions at sea and subsequently check that such action is having the desired effect.
44. In clear weather, whenever possible, the officer in charge of the navigational watch shall carry out radar practice.

Restricted visibility

45. When restricted visibility is encountered or expected, the first responsibility of the officer of the watch is to comply with the relevant rules of the International Regulations for Preventing Collisions at Sea, with particular regard to the sounding of fog signals, proceeding at a safe speed and having the engines ready for immediate manœuvre. In addition, the officer in charge of the navigational watch shall:
   1. inform the master,
   2. post a proper look-out,
   3. exhibit navigation lights, and
   4. operate and use the radar.

In hours of darkness

46. The master and the officer in charge of the navigational watch when arranging look-out duty shall have due regard to the bridge equipment and navigational aids available for use, their limitations; procedures and safeguards implemented.

Coastal and congested waters

47. The largest scale chart on board, suitable for the area and corrected with the latest available information, shall be used. Fixes shall be taken at frequent intervals, and shall be carried out by more than one method whenever circumstances allow.

48. The officer in charge of the navigational watch shall positively identify all relevant navigation marks.

Navigation with pilot on board

49. Despite the duties and obligations of pilots, their presence on board does not relieve the master or officer in charge of the watch from their duties and obligations for the safety of the ship. The master and the pilot shall exchange information regarding navigation procedures, local conditions and the ship's characteristics. The master and/or the officer in charge of the navigational watch shall co-operate closely with the pilot and maintain an accurate check on the ship's position and movement.

50. If in any doubt as to the pilot's actions or intentions, the officer in charge of the navigational watch shall seek clarification from the pilot and, if doubt still exists, shall notify the master immediately and take whatever action is necessary before the master arrives.

Ship at anchor

51. If the master considers it necessary, a continuous navigational watch shall be maintained at anchor. While at anchor, the officer in charge of the navigational watch shall:

1. determine and plot the ship's position on the appropriate chart as soon as practicable;
2. when circumstances permit, check at sufficiently frequent intervals whether the ship is remaining securely at anchor by taking bearings of fixed navigation marks or readily identifiable shore objects;
3. ensure that proper look-out is maintained;
4. ensure that inspection rounds of the ship are made periodically;
5. observe meteorological and tidal conditions and the state of the sea;
6. notify the master and undertake all necessary measures if the ship drags anchor;
7. ensure that the state of readiness of the main engines and other machinery is in accordance with the master's instructions;
8. if visibility deteriorates, notify the master;

   9. ensure that the ship exhibits the appropriate lights and shapes and that appropriate sound signals are made in accordance with all applicable regulations; and

10. take measures to protect the environment from pollution by the ship and comply with applicable pollution regulations.

STCW Code

Part B

Recommended Guidance Regarding Provisions of the STCW Convention and its Annex Section B-VIII/2

Guidance regarding watchkeeping arrangements and principles to be observed

1. The following operational guidance should be taken into account by companies, masters and watchkeeping officers.

Part 3.1: Guidance on Keeping a Navigational Watch

Introduction

2. Particular guidance may be necessary for special types of ships as well as for ships carrying hazardous, dangerous, toxic or highly flammable cargoes. The master should provide this operational guidance as appropriate.

3. It is essential that officers in charge of the navigational watch appreciate that the efficient performance of their duties is necessary in the interests of the safety of life and property at sea and of preventing pollution of the marine environment.

Bridge resource management

4. Companies should issue guidance on proper bridge procedures, and promote the use of checklists appropriate to each ship taking into account national and international guidance.

5. Companies should also issue guidance to masters and officers in charge of the navigational watch on each ship concerning the need for continuously reassessing how bridge-watch resources are being allocated and used, based on bridge resource management principles such as the following:

1. a sufficient number of qualified individuals should be on watch to ensure all duties can be performed effectively;

2. all members of the navigational watch should be appropriately qualified and fit to perform their duties efficiently and effectively or the officer in charge of the navigational watch should take into account any limitation in qualifications or fitness of the individuals available when making navigational and operational decisions;

3. duties should be clearly and unambiguously assigned to specific individuals, who should confirm that they understand their responsibilities;

4. tasks should be performed according to a clear order of priority;

5. no member of the navigational watch should be assigned more duties or more difficult tasks than can be performed effectively;

6. individuals should be assigned at all times to locations at which they can most efficiently and effectively perform their duties, and individuals should be reassigned to other locations as circumstances may require;

7. members of the navigational watch should not be assigned to different duties, tasks or locations until the officer in charge of the watch is certain that the adjustment can be accomplished efficiently and effectively;

8. instruments and equipment considered necessary for effective performance of duties should be readily available to appropriate members in charge of the navigational watch;

9. communications among members of the watch should be clear, immediate, reliable and relevant to the business at hand;

10. non-essential activity and distractions should be avoided, suppressed or removed;

11. all bridge equipment should be operating properly and if not, the officer in charge of the navigational watch should take into account any malfunction which may exist in making operational decisions;

12. all essential information should be collected, processed and interpreted, and made conveniently available to those who require it for the performance of their duties;

13. non-essential materials should not be placed on the bridge or any work surface; and

14. members of the navigational watch should at all times be prepared to respond efficiently and effectively to changes in circumstances.

Change of bearing with change of range

Closest approach miles	Range in nautical miles														
	1	2	3	4	5	6	7	8	9	10	11	12	13	14	15
0.25		7.3°	2.4°	1.2°	0.7°	0.5°	0.3°	0.3°	0.2°	0.2°	0.1°	0.1°	0.1°	0.1°	0.0°
0.50		15.5°	4.9°	2.4°	1.5°	0.9°	0.7°	0.5°	0.4°	0.3°	0.3°	0.2°	0.2°	0.2°	0.1°
0.75		26.6°	7.5°	3.7°	2.2°	1.4°	0.8°	0.6°	0.5°	0.4°	0.3°	0.3°	0.2°	0.2°	0.1°
1.0		60.0°	10.5°	5.0°	3.0°	1.9°	1.4°	1.0°	0.9°	0.7°	0.6°	0.4°	0.3°	0.3°	0.3°
1.5			18.6°	8.0°	4.5°	3.0°	2.1°	1.5°	1.2°	1.0°	0.8°	0.6°	0.5°	0.4°	0.4°
2.0			48.2°	11.8°	6.4°	4.1°	2.9°	2.1°	1.6°	1.3°	1.0°	0.8°	0.7°	0.6°	0.5°
2.5				17.7°	8.7°	5.4°	3.7°	2.7°	2.0°	1.7°	1.4°	1.1°	0.9°	0.8°	0.7°

Example. For a closest approach of 1.0 mile, the change of bearing as the range decreases from 12 to 9 miles = 0.4° + 0.6° + 0.7° = 1.7°.

A Guide to the Collision Avoidance Rules. DOI: 10.1016/B978-0-08-097170-4.00010-6

Manœuvres to avoid collision

The Rules of Part B Section II require power-driven vessels in sight of one another when in a meeting situation to turn to starboard and, when in a crossing situation, to avoid crossing ahead of a vessel on the starboard side and to avoid turning to port for a vessel on the port side. Rule 19(d), for vessels not in sight of one another, requires that alterations of course to avoid a close quarters situation should not be to port for a vessel forward of the beam and should not be towards a vessel abeam or abaft the beam.

These Rules effectively require that alterations of course should normally be to starboard for a vessel forward of the beam and on the port quarter and to port for a vessel on the starboard quarter. However, no guidance is given in the Rules as to the magnitude of such alterations, apart from the requirement that they should be substantial and large enough to be readily apparent to another vessel observing visually or by radar (Rule 8(b) and (c)).

If each of two vessels, approaching so as to involve risk of collision in a meeting or crossing situation, detects the other forward of the beam and alters course to starboard the action of one vessel will usually complement the action of the other. Each vessel would be taking action which would cause the line of sight to rotate in an anti-clockwise direction, i.e., cause the compass bearing to decrease. However, the effectiveness of helm action will obviously not continue to increase indefinitely with the angle through which the vessel turns. There will always be an optimum value beyond which the effect on the nearest approach will be reduced. An alteration to starboard equal to twice the angle between the bearing of the other ship and the port beam will have no effect on the distance of nearest approach.

The helm action which will initially be most effective in causing anti-clockwise rotation of the line of sight is an alteration to port or starboard to bring the other vessel abeam to port. For this purpose alterations of course would need to be to starboard for a vessel forward of the beam and to port for a vessel abaft the beam.

If a vessel is approaching from the vicinity of the port beam it will not be possible to take helm action which would complement the probable action of the other vessel and an alteration to port could be dangerous if a vessel approaching from the port quarter keeps her course and speed. When a vessel approaching from the port beam or port quarter fails to keep out of the way an alteration of course to starboard to bring the other vessel astern, or nearly astern, would probably be the safest form of avoiding action. Such action would be in accordance with Rule 19(d) for vessels not in sight of one another.

A working party set up by the Royal Institute of Navigation in 1970 to consider possible changes to the Regulations for Preventing Collisions at Sea suggested that a manœuvring diagram would be of value as a guide to mariners. The diagram illustrated

A Guide to the Collision Avoidance Rules. DOI: 10.1016/B978-0-08-097170-4.00011-8

in this section is almost identical with the one recommended by a majority of the working party. A minor modification has been made to avoid conflict with Rule 19(d).

The diagram is restricted to course alterations but advice concerning changes in speed is given in the accompanying notes compiled by the working party. The diagram is intended primarily for use when avoiding a vessel detected by radar but not in sight. Smaller alterations will usually be sufficient when avoiding a vessel in sight which is approaching from ahead or from the starboard bow.

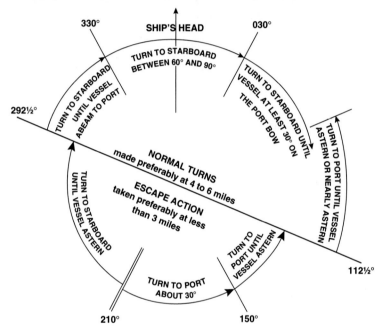

Course Alteration Diagram intended primarily for use in avoiding a vessel detected by radar but not in sight. Journal of the Institute of Navigation, Vol. 25, Page 105

Resumption of course

After turning to starboard for a vessel on the starboard side keep the vessel to port when resuming course.

Escape action

A vessel approaching from the port beam and astern sector can normally be expected to take early avoiding action. The suggested turns are recommended for use when such a vessel fails to keep out of the way. As an alteration of course to put the bearing astern may not complement subsequent action by the other vessel it is recommended that further turns be made to keep the vessel astern until she is well clear.

Speed Changes in Restricted Visibility

Reductions of speed

A vessel is permitted to reduce speed or stop at any time and such action is recommended when the *compass* bearing of a vessel on the port bow is gradually changing in a clockwise direction (increasing). A reduction of speed should be made as an alternative to, and not in conjunction with, the suggested turns to starboard for avoiding a vessel either on the port bow or ahead. Normal speed should be resumed if it becomes apparent that a vessel on the port side has either subsequently turned to starboard in order to pass astern or stopped.

Increases of speed

It will sometimes be advantageous to increase speed if this is possible within the limitations of the requirement to proceed at a safe speed. An increase of speed may be appropriate when the vessel to be avoided is astern, or on the port quarter, or near the port beam, either initially or after taking the helm action indicated in the diagram.

Limitations

The presence of other vessels and/or lack of sea room may impose limitations on the manœuvres which can be made, but it should be kept in mind that small changes of course and/or speed are unlikely to be detected by radar.

CAUTION. IT IS ESSENTIAL TO ENSURE THAT ANY ACTION TAKEN IS HAVING THE DESIRED EFFECT. If not the recommended turns can normally be applied successively for newly developed collision situations with the same vessel.

Manœuvring information

Typical crash stop distances for ships of various types

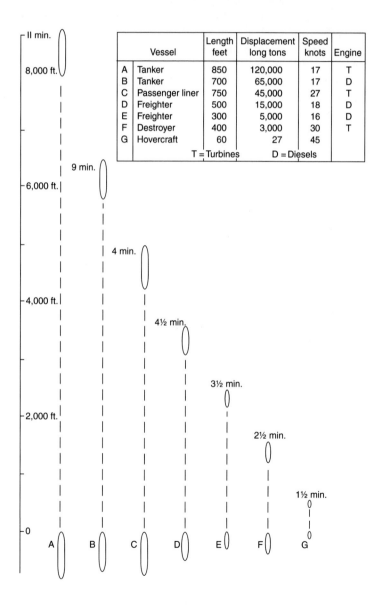

	Vessel	Length feet	Displacement long tons	Speed knots	Engine
A	Tanker	850	120,000	17	T
B	Tanker	700	65,000	17	D
C	Passenger liner	750	45,000	27	T
D	Freighter	500	15,000	18	D
E	Freighter	300	5,000	16	D
F	Destroyer	400	3,000	30	T
G	Hovercraft	60		27	45

T = Turbines   D = Diesels

A Guide to the Collision Avoidance Rules. DOI: 10.1016/B978-0-08-097170-4.00012-X

Turning circles

A typical turning circle for a merchant vessel turning at full speed with rudder hard over is shown below. The following points should be noted:

1. *Pivot point.* This is the point about which the vessel turns. It is usually about $^1/_3$ of the vessel's length from the stem when going ahead.
2. *Path traced out by pivot point.* The vessel turns slowly to begin with due to her initial momentum so the path is not a perfect circle. The pivot point is likely to be displaced initially away from the side to which the vessel is turning due to the pressure against the rudder.
3. *Path traced out by stern.* The vessel must be expected to move through at least 2 ship lengths before the stern clears the original path.
4. *Advance.* This is usually between 3 and 5 ship lengths. The time taken to turn through 90° would be between 2 and 3 minutes for most vessels.

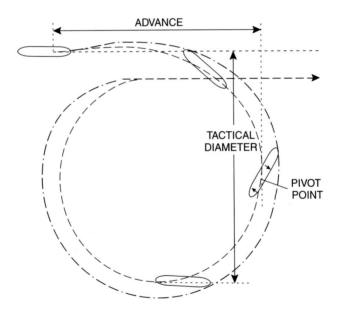

5. *Loss of speed.* By the time the vessel has turned through 90° she will probably have reduced her speed by about $^1/_3$, and after 180° by approximately ½.
6. *Tactical diameter.* This is usually slightly greater than the advance.
7. *Complete circle.* The time taken for a complete turn is likely to be between 5 and 10 minutes. The vessel will probably end up inside her original track.
8. *Variation.* A vessel which is fully loaded will have a larger turning circle and will take a greater time than when she is light. A right handed single screw ship may be expected to turn better to port than to starboard. The effect of wind will vary according to the type of vessel.

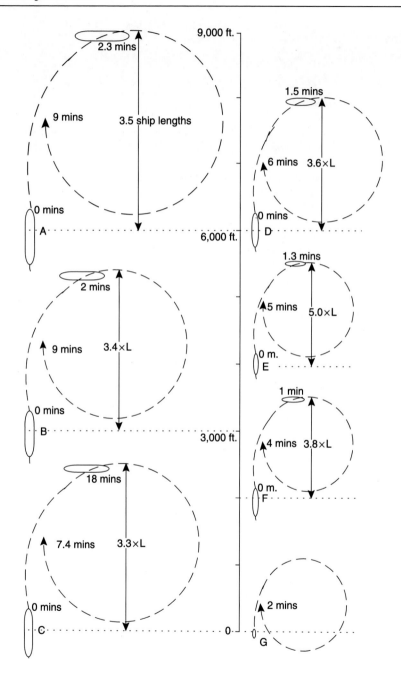

List of ships

Key to abbreviations

L.R.	Lloyd's List Law Report. (Volume number precedes abbreviation and page number follows)
Asp.	Aspinall's Reports of Maritime Cases.
F.I., R.	Report of Court of Formal Investigation.
USCGI	United States Coast Guard Investigation.
U.S.C.	United States Court Case.
N.I.	Netherlands Inquiry.
L.I.	Liberian Inquiry.
M.A.I.B.	Marine Accident Investigation Branch Report.

Aegean Captain (Atlantic Empress) 1980 L.I.	23
Alcoa Rambler (Norefjord) 1949 82 L.R. 359	75
Almizar (John C. Pappas) 1969 1 L.R. 1	25
Andulo (Statue of Liberty) 1970 2 L.R. 159 1971 2 L.R. 277	30
Angelic Spirit (Y Mariner) 1994 2 L.R. 595	28, 82–3
Anna Salen (Thorshovdi) 1954 1 L.R. 475	46, 102
Anneliese (Arietta S Livanos) 1970 1 L.R. 36	15
Antonio Carlos (Bovenkerk) 1973 1 L.R. 70	14
Aquarius (Atlantic Hope) 1978 U.S.C.	54
Aras (Oakmore) 1907 10 Asp. 358	100, 102
Arietta S Livanos (Anneliese) 1970 1 L.R. 36	15
Aristos (Linde) 1969 2 L.R. 568	102
Ashton (King Stephen) 1905 10 Asp. 88	77
Asian Energy (Century Dawn) 1993 1 L.R. 138	57–8
Atlantic Empress (Aegean Captain) 1980 L.I.	23
Atlantic Hope (Aquarius) 1978 U.S.C.	54
Atys (Siena) 1963 N.I.	31
Auriga (Manuel Campos) 1977 1 L.R. 384	71
Baines Hawkins (Moliere) 1893 7 Asp. 364	69
Banshee (Kildare) 1887 6 Asp. 221	26
Billings Victory (Warren Chase) 1949 82 L.R. 877	37
Boulgaria (Hagen) 1973 1 L.R. 257	20
Bovenkerk (Antonio Carlos) 1973 1 L.R. 70	14
Bremen (British Grenadier) 1931 40 L.R. 177	100
British Aviator (Crystal Jewel) 1964 2 L.R. 403 1965 1 L.R. 271	35, 38
British Engineer (Karanan) 1945 78 L.R. 31	74–5
British Grenadier (Bremen) 1931 40 L.R. 177	100

A Guide to the Collision Avoidance Rules. DOI: 10.1016/B978-0-08-097170-4.00013-1

Index

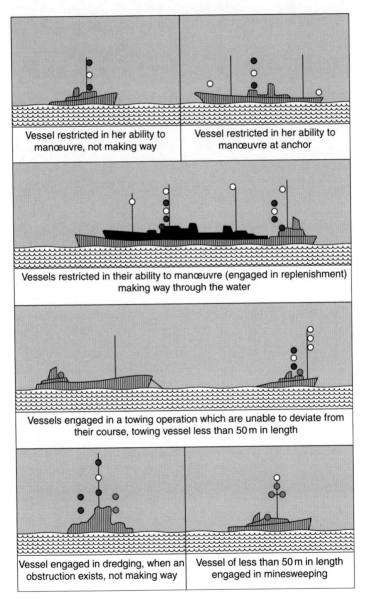

Note: The white masthead lights should be placed above and clear of all other lights.

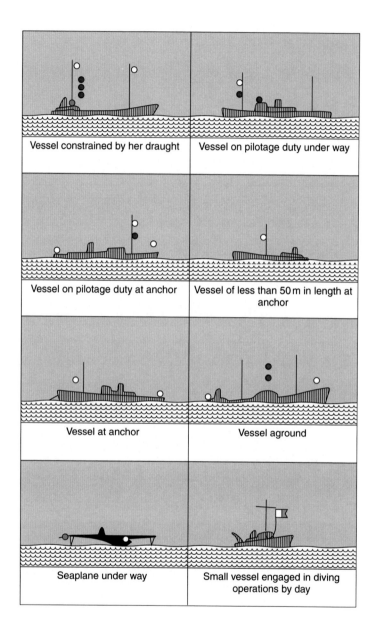

Vessel constrained by her draught

Vessel on pilotage duty under way

Vessel on pilotage duty at anchor

Vessel of less than 50 m in length at anchor

Vessel at anchor

Vessel aground

Seaplane under way

Small vessel engaged in diving operations by day

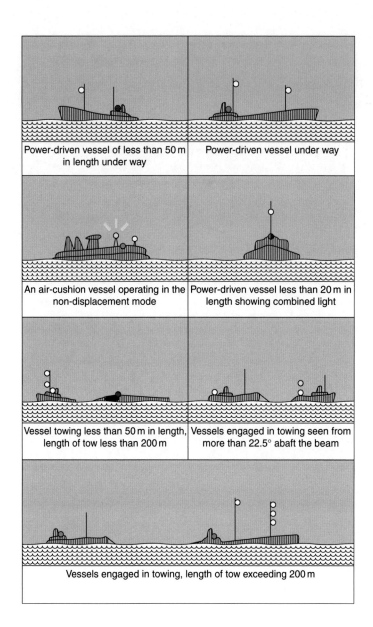

Power-driven vessel of less than 50 m in length under way

Power-driven vessel under way

An air-cushion vessel operating in the non-displacement mode

Power-driven vessel less than 20 m in length showing combined light

Vessel towing less than 50 m in length, length of tow less than 200 m

Vessels engaged in towing seen from more than 22.5° abaft the beam

Vessels engaged in towing, length of tow exceeding 200 m

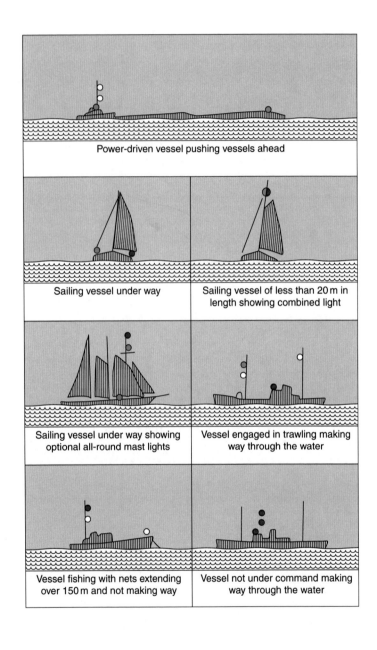

Power-driven vessel pushing vessels ahead

Sailing vessel under way

Sailing vessel of less than 20 m in length showing combined light

Sailing vessel under way showing optional all-round mast lights

Vessel engaged in trawling making way through the water

Vessel fishing with nets extending over 150 m and not making way

Vessel not under command making way through the water